GENDER, PLACE AND THE

Gender, Place and the Labour Market

SARAH JENKINS
University of Wales, Aberystwyth

Routledge
Taylor & Francis Group

LONDON AND NEW YORK

First published 2004 by Ashgate Publishing

Reissued 2018 by Routledge
2 Park Square, Milton Park, Abingdon, Oxon OX14 4RN
711 Third Avenue, New York, NY 10017, USA

Routledge is an imprint of the Taylor & Francis Group, an informa business

First issued in paperback 2018

A Library of Congress record exists under LC control number: 2003065369

Notice:
Product or corporate names may be trademarks or registered trademarks, and are used only for identification and explanation without intent to infringe.

Publisher's Note
The publisher has gone to great lengths to ensure the quality of this reprint but points out that some imperfections in the original copies may be apparent.

Disclaimer
The publisher has made every effort to trace copyright holders and welcomes correspondence from those they have been unable to contact.

ISBN 13: 978-0-815-38917-0 (hbk)
ISBN 13: 978-1-138-61974-6 (pbk)
ISBN 13: 978-1-351-15764-3 (ebk)

Contents

List of Figures

Preface

Sarah Jenkins' research builds on a now long tradition of enquiry into the geography of women's participation in paid employment. And although, since the 1980s, the growing body of innovative feminist research in geography has become increasingly wide-ranging and diverse in theoretical and empirical perspectives and preoccupations (WGSC 1997, McDowell 1998), there has been a continuing concern with the geography of women's employment and the constraints on their labour market participation. Indeed, some of the earliest feminist geography in Britain focused on these themes (Tivers 1982, Lewis 1984). Other research has since addressed key issues including: the gendering of workplace relations; the significance of local, gendered social networks in job search and hiring; the development of local cultures of mothering; and the importance of the diversity of women's social identities to explanations of their labour market behaviour.

However, we still seek richer answers to those early questions concerning the spatial variation in women's participation in paid work. How do rates of participation vary at different spatial scales? How can we account for these variations? Are they simply a product of the geography of economic activity or – more probably – do they arise from complex interconnections between local ways of life, local patterns of gender relations and shifting structures of local employment? Are there spatial variations in practices and beliefs concerning childcare and women's participation in paid work? For example, are there spatial variations in the involvement of men in childcare, in women's views of how children should be cared for and of whether and for what hours mothers should take on paid work? Do such spatial variations only reflect differences of class or economy or, once again, are locally based social relations influential?

These questions are necessarily central to an understanding of the continuing economic inequality between women and men (Women and Equality Unit 2002). And although we have partial answers there still remains a need for more extensive analysis of the spatial variation in women's participation in paid employment and deeper research into local variation in gender relations. Sarah Jenkins' research addresses and contributes to both these key areas.

She begins by exploring the growing sophistication of research on women's labour market participation. She identifies the recent emphasis on the importance of developing theoretically informed local labour market research. Her discussion of recent national trends in women's labour market participation is followed by an analysis of Labour Force Survey statistics that identifies spatial variations in women's participation in paid work at the local authority district level. There is then a fascinating in-depth exploration of the labour market activity of a sample of women interviewed in two areas with strongly contrasting participation levels – Neath Port Talbot and West Dorset. The interviews provide the basis for a subtle and thoughtful analysis of women's labour market decision making in the two areas. The book concludes with a discussion of the policy implications of the research, including a plea for greater practical recognition of the importance of local geographies. The research provides a clear demonstration of the continuing value of research and policy making that recognises the interdependencies of local and national socio-economic processes.

Sophie Bowlby

References

Lewis J (1984) *Post-War Regional Development in Britain – the Role of Women in the Labour Market*, unpublished PhD thesis, Queen Mary College, London University.

McDowell L (1998) *Gender, Identity and Place: Understanding Feminist Geographies*, Cambridge, Polity Press.

Tivers J (1982) *Weekday Spatial Activity Patterns of Women with Young Children*, unpublished PhD thesis, University of London.

Women and Equality Unit (2002) *Key Indicators of Women's Position in Britain*, London, Department of Trade and Industry.

Women and Geography Study Group of the IBG (WGSG) (1997) *Feminist Geographies: Explorations in Diversity and Difference*, Harlow, Longman.

Acknowledgements

Although the name on the cover is mine, *Gender, Place and the Labour Market* has evolved due to the input, support and belief of many people.

Firstly, there are Martin Jones and Mark Goodwin who placed their belief in me at a very early stage and who have provided nothing but encouragement along the whole way. Thank you for allowing me to develop and for looking over endless earlier versions of my work. A special thanks to Martin for never failing to answer my continuous questioning and for allowing me to 'sound off' during times of panic. Thanks also go to Luke Desforges for encouraging an early interest in this subject area, and for his input during the later stages of this project. Thank you also to Sophie Bowlby for your kind words, support and encouragement with this project. My gratitude goes to the Institute of Geography and Earth Sciences at the University of Wales, Aberystwyth, who provided financial support in order to help me undertake the research for this project.

Although the names of all the women who spent time chatting to me must remain anonymous, I must thank each and every one of them for opening their lives to me, inviting me into their homes and for the endless cups of tea. From listening to these women I learnt a lot, laughed a lot, ate a lot of biscuits and in some cases wanted to shed a tear. However, without their honesty, this book would not have been possible.

On a personal note, massive thanks go to mum, dad and Mark who have allowed me to follow my own path, continually supporting me the whole way. I apologise to mum and dad for taking over their office and hogging the computer during the final stages of this project – but I hope it was worth it! Mark, I have thus far avoided that job in McDonald's, but it is only a matter of time....

I am indebted to Blondie, Carys, Catho, Chrissie, Emma, Gwawr, Karen, Kelly, Linda, Mari, Philippa, Rachael, Rhi, Rhian, Rosie, Sara and Tina who have supported me through my endless years of being a student, backed my mad ideas and maintained my sanity – thanks girls, a wholehearted hug to you all!

The following figures appear in this book with permission:

Figure 4.3 reprinted from Women and Geography Study Group of the Institute of British Geographers (1984) *Geography and Gender: An introduction to feminist geography* with permission from the Women and Geography Study Group.

Figure 4.5 reprinted from Cooke, P. (1983) *Theories of Planning and Spatial Development* with permission from the author.

Figure 5.1 reprinted from Sayer, A. (1992) *Method in Social Science: A realistic approach* with permission from Thomson Publishing Services.

The following substantial quotes appear in this book with permission:

Page 87, reprinted with permission of Sage Publications Ltd from Bennett and McCoshan (1993) *Enterprise and Human Resource Development: Local capacity building.*

Page 72, reprinted with permission of Thomson Publishing Services from Massey, D. & Meegan, R. (1982) *The Anatomy of Job Loss: The how, why and where of employment decline.*

Every effort has been made to obtain permission to reproduce copyright material. If any proper acknowledgement has not been made, I would invite copyright holders to inform me of the oversight.

List of Abbreviations

DfEE	Department for Employment and Education
	Now DfES (Department for Education and Skills)
DTI	Department of Trade and Industry
ESRC	Economic and Social Research Council
EU	European Union
GDP	Gross Domestic Product
ILO	International Labour Organisation
IoD	Institute of Directors
LAD	Local Authority District
LFS	Labour Force Survey
MEE	Motherhood Employment Effect (see Duncan and Savage 2002)
NPT CBC	Neath Port Talbot County Borough Council
ONS	Office for National Statistics
UK	United Kingdom
USA	United States of America
WAG	Welsh Assembly Government
WDDC	West Dorset District Council
WEU	Women and Equality Unit
WGSG	Women and Geography Study Group of the Institute of British Geographers
WPT	The Work and Parents Taskforce
WU	Women's Unit

Chapter 1

Introduction:
Women, Work and Home in the Twenty-first Century

Introduction

> 'When Lisa Gordon announced last week that she was giving up her six-figure salary to spend more time with her children, her story was widely reported in the tabloid press, but given an all-too-familiar spin. High-flying career woman, tried to have it all, discovered she couldn't, came to her senses and returned to the domestic fold.
>
> The 36-year-old corporate affairs director of Chrysalis Records, who earns £133,000 a year (she was on £336,000 until she went part-time last year), "faced the classic career woman's dilemma", it was reported. "From now on, however, school runs will take over from share prices and homework will supersede boardroom battles," says the Daily Mail.'
>
> (The Guardian, Monday 2nd December 2002)

Can women really have it all? In fact, do women really want it all? Daily life in early twentieth-century society largely revolved around very clearly defined patterns amongst men and women: men overwhelmingly travelled to work in order to provide the household with an income, whilst women stayed in the home performing the domestic duties (Mackenzie 1986). Woman's role was seen firstly as that of wife and mother, and secondly as a provider of unskilled, low paid labour between leaving school and starting a family (Massey 1994). In the past fifty years there have been dramatic changes in the attitudes and perceptions towards woman's role within society. Changing structures in society such as the introduction of reliable contraception, new patterns of consumption and a restructuring of the labour market through the introduction and implementation of new technologies, have meant that more women that ever before are taking formal paid employment outside of the family home.

Entry by women into the labour market, however, has not been accompanied by any re-negotiation of their domestic role and women are still responsible for the majority of domestic tasks, regardless of their involvement in paid employment.

Little et al. (1988) and Tivers (1977) believe that this dual role places severe constraints on women's labour market activity due to the restriction that domestic duties place on the amount of time a woman has available to undertake paid employment. The UK Government is keen to promote initiatives which encourage women into the workforce, and much time is spent advocating the economic and social benefits gained from being a working mother. However, despite the Government efforts, some women maintain the opinion that it is their role to be the primary carer for their children and therefore insist that any paid formal employment must not interfere with that responsibility. A woman's own perception or interpretation of her role is as important a constraint on activities as any externally imposed restriction. Practical and tangible support can be given such as childcare (Tivers 1986, Arber and Gilbert 1982, Brannen and Moss 1991). However, childcare is not the only barrier to women, for as McDowell (1997) argues, women's working lives still differ from men's even amongst the highly educated, well-paid and childless women. Evidence of this is found within gendered cultures which form both within and outside of the workplace (see also Deal 1988, McDowell 1997, Orenstein 2000, Turner 2000).

The current UK Labour government is keen to promote *choices* for women, providing them with the opportunity to have quality time with their child(ren) at the same time as fulfilling their domestic role, whilst also allowing them to maintain a valued career position within the new economy. As the current Minister for Women puts it:

> 'I think ... what Government has to do, and increasingly what employers have to do, is to enable different people to make different choices about how they balance work and family, work and the rest of their lives, at different stages in their increasingly long lives. For families with children that means Government mustn't in any way dictate to families about how they balance earning and caring, paid and unpaid work. Government instead has to create the supportive environment, the provision of public services, the right employment regulatory framework to ensure that parents can make choices about how they balance earning a living and caring for their children in the ways that will suit them and their families best. And of course that commitment to greater choice, real choice for more and more parents, is entirely in line with the most fundamental values of our Government.'
>
> (Hewitt 2002)

Given this context, and through an analysis of national, regional and local scale data, this book explores the *geography* of women's participation in the UK labour market, and investigates the factors, which both influence women's decision-making process and contribute to the formation of their perceived societal role. Exploring labour market geographies through various scales of analysis, this book starts with an investigation of women's formal labour market participation at the national scale including an examination of government claims regarding women's labour market position within the UK and their explanations for that. I shall then move to the meso scale with an exploration of women's labour market activity at local authority district level, identifying that in fact there is a large spatial variation of women's economic activity across the UK, which is not evident at the national scale. In order to investigate an explanation for the spatial variation, I shall present the findings of discussions with individual women about the influential factors in their decision-making process regarding their participation or non-participation in the formal labour market. This book will demonstrate the importance of local social processes in addition to the role of economic factors in making labour market geographies. Furthermore, this book will explore exactly what *choices* the Government claims it provides, and what *choices* individual women feel they have when it comes to negotiating their everyday lives.

The Structure of This Book

Chapter 2 situates my research within a wider context of academic debates on labour markets and their geography. I follow the development of, and draw together two very distinct sets of literature; labour market segmentation theory (see Peck 1996a) and the development of feminist debates within geography. Initially there was very little overlap between the two debates, and they developed in parallel to each other, but chapter 2 argues that within contemporary academic debate they do need to come together if we are to gain a fuller understanding of women's labour market participation (see Hakim 2000).

Chapter 3 examines women's position within the labour market from a national context. After positioning women's current UK labour market activity rates with respect to the rest of Europe, the chapter maps how women's participation within the formal labour market has changed over the last fifty years, notably increasing, as men's economic activity slowly decreases. This chapter explores the current demands and expectations on women as the UK enters a new 'knowledge based economy' (see Hakim 2000) where women's transferable skills are vital in taking the economy forward. Therefore, in order to utilise women's skills, the Government is trying to encourage mothers back into the workforce by establishing a better work:life balance through the promotion of initiatives such as flexible working, the National Childcare Strategy and increasing maternity and paternity rights. This chapter looks at UK national policy and considers what this means for individual women.

Chapter 4 moves beyond the national level to the meso scale and establishes that national level data is in fact not representative of what women are actually experiencing. This chapter demonstrates that there is a wide variation in women's economic activity rates across the UK not recognised in national level analysis.

Looking at local authority level data and through an examination of previous academic literatures, this chapter critiques the ways in which typologies have been used to explain why women's economic activity in the UK ranges from 93 per cent in West Dorset to 40 per cent in Newham.

Having established that a spatial variation of women's economic activity exists across the UK, chapter 5 highlights the methodological considerations required to move beyond a national level analysis and investigate why the variation in women's labour market activity exists. After a discussion on the benefits of using both extensive and intensive research techniques, this chapter describes how I used a combination of the two in order to gain a fuller understanding of women's labour market activity. This chapter goes on to introduce the two case study areas for this book, providing an economic geography of each area as well as an analysis of an extensive survey carried out on mothers in each of the areas, exploring their current situation regarding labour market participation.

Chapters 6 and 7 present the results of intensive research in the two case study areas, which explores in-depth the reasons why women have taken their respective decisions regarding whether to work or not work, once they have had children. Chapter 6 presents the results of in-depth interviews carried out with women at a local level in Neath Port Talbot and West Dorset, who are currently trying to combine motherhood with paid work in the formal labour market, whilst chapter 7 focuses on those women who have withdrawn from the labour market and have 'chosen' to stay at home to be a full-time mother. The interviews asked the women about their chosen 'career path' and enquired why they had *chosen* to take that particular route. Both chapters establish that rather than a single barrier existing to prevent women entering or not entering the labour market, it is in fact a complex combination of interdependencies which overlap to influence a woman's decision on whether or not she wants, and is able, to enter the formal labour market.

Having explored the issues that are important to individual women at the local level, chapter 8 draws together all the findings and explores the issue of conceptualising spatial variation in women's formal labour market participation. Drawing out the similarities and differences between the responses from the women in the two localities, this chapter highlights the impact of local social relations on the implementation and conduct of national level policies. This chapter gives prominence to the fact that some of the major influencing factors in a woman's decision-making process are dominated by locally determined social factors, rather than the broader economic factors which form the basis of nationally driven Government policy. As a result, this chapter offers suggestions on policy reform which must be considered if the Government is to offer genuine *choices* to women.

Chapter 2

Gendering the Labour Market

Introduction

It has long been argued by labour market theorists that men and women operate in different labour market contexts (see Barron and Norris 1976, Rubery 1988, Picchio 1992, Peck 1996a). In this chapter I will attempt to trace the development of two sets of literature, which have explored women's position either within or outside of the labour market: firstly, labour market segmentation theory, which has explored women's labour market activity from its understanding of the ways in which labour markets operate; secondly, the development of literature on gender within geography which acknowledged that women operate primarily from the home, and consequently it was in fact the home which determined labour market activity not vice versa. Although these two areas of academic literature have been explored and developed separately, there are times when themes within the two literatures come together in their discussion of women's position either in the labour market or within the home. I will attempt to draw attention not only to where the links are in the two literatures, but also identify lacunae in terms of trying to explain women's position within the labour market. Both the progression of labour market theory and the development of gender literature within geography have diversified to provide a multifaceted literature base. In order to unpack the different literatures and to analyse their different contributions, I will deploy a chronological and thematic approach. By doing this, I will demonstrate their complementarities. Additionally, for the purpose of this review I will concentrate specifically on those arguments that I feel have contributed towards explaining women's decision making, regarding their subsequent role within or outside the labour market.

Separation of Men and Women

Initial literature from both labour market theorists and feminist geographers treated men and women almost as two separate categories. The main goal of both first and second generation segmentation theory and the initial impetus of feminist thought within geography, was to highlight the fact that differences existed in the type of work women undertook in comparison to men. Labour market theorists argued these were due to the level of control which the labour market had and the recognition by employers that women had an alternative role outside of the workplace which would impinge on their commitment.   In contrast, feminist geographers argued that it was the alternative role of homemaker and childrearer which dictated a woman's position in or out of the labour market, which was reinforced through systems of patriarchy which existed in order to maintain women's position as subordinate to men's.

First Generation Segmentation

Theories of labour market segmentation have as their basis, the notion of the dual labour market initially developed by Doeringer and Piore in the 1960s (Peck 1989a, 120; also see Doeringer and Piore 1971).   Doeringer and Piore's work represents the first generation of segmentation theory, with their development of the concepts of primary and secondary sectors within the labour market.   The primary sector contains jobs, which offer relatively high wages, and provides stable and secure employment to workers who can expect to enjoy some form of career progression through an internal labour market.   During their research, based in the United States, Doeringer and Piore found that jobs within the primary sector were primarily occupied by white men.   The secondary sector contains the market's least desirable jobs, described by Kreckel as jobs for workers with 'everyman's qualifications' (1980, 542), and consists of poor wages, poor working conditions and few opportunities for promotion.   The majority of employment in this sector is associated with small firms and is, therefore, prone to strong competitive pressures, staff turnover is high and the threat of unemployment is constant.   In the industrial West, the low status jobs in the secondary sector tend to be filled by ethnic minority workers, women, the disabled and young people (Peck 1996a).   According to Barron and Norris there seem to be five main attributes that make a particular social group or category a likely source of secondary workers; dispensability, clearly visible social difference, little interest in acquiring training, low economism and lack of solidarity (1976, 53).   Barron and Norris fit women into each of these criteria due to their relationship towards their family, and the belief shared by many women that careerism does not accord with their place in the family after marriage.   Similarly, Craig et al. (1985) also see women as secondary workers believing that they are prepared to take low-paid and unrewarding jobs, not because of a weak attachment to work and limited income needs, but because of the supplementary importance of their earnings to family living standards and the constraints on their choice of job.

The secondary sector provides a great deal of the flexibility favoured by the economic system, as increases in output required in the primary sector at the peak of the business cycle can be achieved by recruiting secondary workers on a temporary basis or by subcontracting from the primary to the secondary (Peck 1996, 51).   This flexibility in the secondary sector is not nearly so readily available in the primary sector, and provides just one of the reasons why women are restricted from entering the latter.   Many labour market theorists have agreed that women's access to the stable sector of the labour market has been restricted, explaining that it is because they are perceived to have an 'alternative' role outside the waged labour market to which they may be drawn (see Peck 1989a, Gordon et al. 1982, Barron and Norris 1976). It is the recognition of women's alternative role: that of homemaker and childrearer rather than a primary worker, which sparked a debate amongst geographers about the extent to which the 'predetermined societal role' has an affect on women's lives. Even for those women who do not conform to the stereotypical work pattern, the very expectation amongst employers that they will, or may conform to it serves to restrict their access to primary jobs (Peck 1989a, 130).

First generation segmentation theory, therefore, acknowledged that women occupy a different sector of the labour market to men with initial explanations for this given as the recognition of women's alternative roles. Moreover, Rubery (1988) and Picchio (1992) both believe that women are likely to remain trapped in the most unstable segments of the labour market until there is a change in their (real and perceived) position within the household division of labour, or until these household 'responsibilities' become more compatible with primary sector employment (see Peck 1989a, 130).

Explanations for women's subordinate position within the labour market have not only come from labour market theorists, but have also come from feminist geographers working in this field. Massey and Meegan (1982) suggest two factors which differentiate male and female labour. Firstly there is the ideological factor, including a belief about woman's capability and the level of skill of the work undertaken by women (which may be underestimated). Secondly, Massey and Meegan move beyond the labour market and identify the sexual division of labour outside the workplace, particularly the greater responsibility of women in domestic work and the restrictions on women's time and daily mobility. Hanson and Pratt (1995) later argue that the positioning of women in the secondary sector can be explained by a combination of both women's tendency to arrange their paid employment around the schedules of their husbands and children, and the sexist practices of male employers and employees. They suggest that male employers may be reluctant to hire women for the most prized jobs because of gendered stereotypes and worries about complaints from male employees. I feel that at this point it is worth noting that feminist geographers primarily considered the 'home role' to be the main influence in a women's decision to enter the labour market, and the consideration of the 'rules' of the labour market are a secondary influencing factor for them.

The Geography of Women: The Initial Argument

In the mid to late 1970s feminist geographers sought to demonstrate that due to gender inequalities, women's access to opportunities was not equal to that of men (Bowlby et al. 1989, 160). Patterns of accessibility to transport and other services, such as childcare, conspire to constrain women's access to paid employment and other urban resources. This first phase of feminist geography initiated a recognition that women as individuals or as a class, exist under different conditions and constraints than do men. In some senses this was a similar point made by those involved in the development of first generation segmentation theory. At first such work was broad in its focus, but as early as 1982, Porter acknowledged that women's position in the labour market can not be fully understood without taking into account the complex connections between waged labour, sexual divisions, and the structure and ideology of the family in capitalist society (Porter 1982, 117). Thus, the key concern of geographers at this time was to document the extent to which women were systematically disadvantaged in many areas of life by the sets of assumptions made about 'woman's place' and by the resulting material constraints on their activities (Bowlby et al. 1989, 158). Zelinsky et al. (1982), for instance, argued that by taking gender relations seriously one could produce a different understanding of societal composition, a more advanced evaluation of regional policy, and a far better understanding of the organisation and reorganisation of our national economic space, including the structure of labour markets.

Gender Role Constraint

Tivers (1977) claimed that the gender role constraint is the one that underlies all other influences on women's activity patterns. Their constraint is underpinned by the social expectation that women's main activities should be those of family care and household maintenance, and the assumption that women will interrupt their working lives to care for children and elderly relatives (Women and Geography Study Group 1988, Bowlby et al.1989). Chodorow (1992) has attempted to explain where the societal assumption of women's primary role has evolved from, arguing that because women are themselves mothered by women, they grow up with the relations, capacities and needs, and psychological definition of self-in relationship, which commit them to mothering. Men, because they are mothered by women, do not. Chodorow suggests therefore, that women mother daughters who, when they become women, mother (1992, 168). Massey describes how the identities of woman and the 'home-place' are intimately tied up with each other (1994, 180) for, as Barron and Norris (1976) suggest, it is often said that a woman's 'real' place is in the home with her family and that her husband is, or should be, the main source of income. Women's perceived traditional role as homemakers and child-rearers, therefore, automatically places them at a disadvantage when participating in paid employment. The constraints placed on women by their domestic responsibilities compel them to state a preference for part-time work, a tendency to leave and return to work as family needs dictate and an inability to work overtime and move location (Cockburn 1988). Many employers assume that women in general have lower income needs and lesser attachment to their job than men, even where in individual cases this assumption may not be applicable (Craig et al. 1985, 60). These sets of literatures demonstrate the extent to which predetermined expectations can potentially constrain women from entering into, and progressing within, the formal labour market.

Patriarchy

According to Walby (1986), gender inequality cannot be understood without the concept of patriarchy. The separation of the primary and secondary workforces as described in labour market theory allowed men to maintain a level of superiority over women in a system which according to Richards, is made 'by men for men' (1988, 5). Patriarchy has been defined as a set of social relations between men, which, although hierarchal, establishes an interdependence and solidarity between them which allows them to dominate women, maintaining that women's interests are subordinate to the interests of men (see Women and Geography Study Group 1984 and Weedon 1987). Patriarchy can be evident in many forms including the sets of rules that allow movement within the labour market. Hartmann (1979) believes that although capital did not create the patriarchal social structures that underpin the marginalised role of women in the labour market, it exploits and thereby perpetuates gender divisions, to the allocation of domestic labour within the home. Walby (1990) presents a system of patriarchy conceptualised as a system of social structures and practices in which men dominate, oppress and exploit women, both within the home and the labour market. According to Walby, the six structures of patriarchy are: firstly, patriarchal production relations within the household. It is through these that woman's household labour is expropriated by her partner, including a position where men generally have access to the family car, leaving the women to rely on slower public transportation; secondly, patriarchal relations within paid work, where a complex form of patriarchal structures exist which exclude women from better forms of work, insuring men remain in the most senior positions in the workplace; thirdly, patriarchal relations in the state, where Walby argues the state has a systematic bias towards patriarchal interests in its policies and actions and where it is the older generation men making policy decisions, including ones solely concerning women; fourthly, male violence, as men enforce themselves upon women to satisfy their own needs. Walby believes that male violence against women is systematically condoned and legitimated by the state's refusal to intervene against it except in exceptional circumstances; fifthly, patriarchal relations in sexuality through the promotion of compulsory heterosexuality and sixthly, patriarchal relations in cultural institutions which is composed of a set of institutions which create the representation of women with a patriarchal gaze (1990, 21). For feminists, the everyday routines traced by women are never unimportant, because the seemingly banal and trivial events of the everyday are bound into the power structures which limit and confine women. The limits on women's everyday activities are structured by what society expects women to be and therefore to do (Rose 1993, 17).

Gender Does Not Just Mean Women

According to Massey, it is not just that geography matters, but it is a gendered geography which matters (1994, 181). Zelinsky et al. (1982) recognise that much of the literature during this first phase of feminist geography was weakened by a tendency to focus on women, rather than emphasising the social divisions between women and men (Bowlby et al. 1989). Massey believes that you cannot discuss one without discussing the other for a feminist geography is (or should be) as much about men as it is about women (1994, 189). It is therefore not possible to look at either men's or women's labour market participation without considering the position of the other, as described by the previous discussion on patriarchy. According to

Zelinsky et al. the researcher 'must view reality stereoscopically, so to speak, through the eyes of both men and women, since to do otherwise is to remain more than half-blind' (1982, 353).

Second Generation Segmentation

The second generation of segmentation theory was developed by radical theorists during the late 1970s, who, in contrast with Doeringer and Piore, emphasised the role of labour market segmentation as a capitalist strategy, which sought to maintain control over the production process (Peck 1989, Gordon 1972). In relation to the dual labour market, Barron and Norris suggest that it is in the interest of the employers to maintain and expand the primary sector and to ensure that instability and low earnings are retained in the secondary sector (1976, 52). Ashton and Maguire (1984) argue that the function of all labour market segments is to restrict the competition for jobs to workers who are already working within the segment. Consequently movement between segments is difficult. Radical theorists argued that monopoly capitalist firms sought to segment their labour forces in the face of declining skill levels through the development of extended hierarchies and exploitation of racial and gender differences. Kreckel notes that as the costs for hiring and/or training of skilled labour are high, large corporations have an interest in minimising their labour turnover and thereby reducing costs. The process of employee deskilling and increasing management control is described by Braverman (1974) as he discusses how the knowledge and skills of workers are constantly incorporated into management functions or into machines, and as a result workers are employed in jobs below their capabilities, which also renders them more easily controlled by management (Thompson 1985, 36). Braverman (1974) integrates his analysis of gender relations; firstly he argues that women take most of the new less-skilled jobs; and secondly, that household tasks are helped with new industrialisation, reducing the amount of labour to be done in the home and therefore releasing women for waged labour. At the same time, by introducing mobility chains, internal training, pension schemes and by exercising control over their workforce, employers try to build up a stable and loyal core personnel. According to Michon (1987), what distinguishes segmentation from mere division is that each segment functions according to different rules. Segmentation theorists of this generation argue that labour markets are 'social constructs, incorporating within them various forms of organisation which both condition their mode of operation and also structure to some extent the actors themselves and determine their behaviour' (Castro et al. 1992, 10). Hierarchies were, therefore, formed within the primary sector itself, creating internal and external labour markets as demonstrated in figure 2.1.

Figure 2.1   Labour market segmentation: the development of internal and external labour markets

Primary Sector					
	Primary Internal		Primary External		
I N T E R N A L	High wages, Advanced working conditions, Strong unionisation, Advanced technology, Autonomous work control, Substantial promotion	e.g. oil, gas, electricity, metallurgy, high-order services	High wages, Good working conditions, Variable unionisation, Advanced technology, Relatively autonomous work control, Little promotion	e.g. Engineering assembly, low order banking, insurance and services	E X T E R N A L
	Secondary Internal		Secondary External		
	Variable wages, Poor working conditions, Low unionisation, Advance technology, Supervised work control, Limited promotion	e.g. engineering components, retailing miscellaneous services	Low wages, Primitive working conditions, Little unionisation, Simple technology, Rigid work rules, Little promotion	e.g. textiles, footwear/ leather, glassware, food processing	
Secondary Sector					

Source: Adapted from Cooke 1983a, 548

Reich et al. (1973) outline that within the primary sector there is a segmentation between 'subordinate' and 'independent' jobs. Subordinate primary jobs are routinised and encourage personality characteristics of dependability, discipline, responsiveness to rules and authority, and acceptance of a firm's goals. According to Reich et al. (1973), both factory and office jobs are present in this segment. In contrast, independent primary jobs require creativity, problem solving, self-initiating characteristics and often, professional standards for work. Voluntary turnover is high and individual motivation and achievements are highly rewarding (p359). Peck suggests that where possible, such divides were deepened by the exploitation of racial and gender cleavages within the workforce; through these practices, employers counteracted tendencies toward solidarism (1996, 53). Ashton and Maguire (1984), in their research of three local labour markets in the UK, discovered that in some larger firms, jobs were part of an elaborate, formally defined internal labour market, where further training was available, and there was the possibility of promotion to a supervisory post. Ashton and Maguire discovered that in smaller firms the informal internal labour markets offered similar opportunities, based on personalised relationships. For example, employers frequently made reference to their 'core workers' and the 'floaters'. The loyal 'core workers' were often paid higher wages than the 'floaters' and were likely to be promoted to supervisory positions (1984, 115). It is the division of the labour market into primary and secondary workforces, and the further separation into internal and external labour markets, which Hakim (1979) terms horizontal (where different types of work are allocated to men and women) and vertical (where men and women participate in the same field of work but women are concentrated in the lower grades) segregation. This division forms the basis of second generation segmentation theory and highlights an emerging gender division within the workplace.

In summary, this initial body of work recognised that men and women occupied different sections of the labour market with men mainly present within the primary sector, which contained well-paid highly skilled jobs and good prospects; and the majority of women occupying the secondary sector which was dominated by poorly paid, low skilled work with no career prospects. These sectors of the labour market were reinforced through patriarchal structures to ensure that the man remained the breadwinner and the woman the homemaker. This book will later explore the impact of these predetermined societal roles.

Recognition of the Dual Role

The second phase of the development of gender within geography highlighted the complex lives which women negotiate between the public, work sphere and the private, domestic sphere. Academics recognised the need to explore the ways in which women have to negotiate each sphere in order to achieve a better understanding of women's lives. After the recognition that more and more women were attempting to combine home and work, geographers Hanson and Pratt (1988a) called for a reconceptualisation of the interdependencies between the two spheres.

The Development of Socialist-Feminist Geography

The second phase of feminist geography ran from the early to late 1980s and according to Bowlby et al. (1989), the analysis of gender through this socialist-feminist stage falls into the following three inter-related categories. First, gender divisions were seen as empirically important as changes in the gender composition of employment and unemployment were seen as a major constituent of dynamic local and regional labour markets, and as a major outcome of the processes of economic restructuring being studied. Secondly, Bowlby et al. argue that it was now increasingly recognised that gender divisions played a major role in influencing the types of change which have taken place in the geography of industry and employment over the post-war period; that gender divisions in the labour market were a major determinant of production and locational change. Finally, the changing gender composition of employment and unemployment was examined in terms of its impact on the composition of local class relations.

In summary, this second phase of feminist geography acknowledged the differences between men and women and began to explore women's lives in an attempt to identify why they differentiate from men. Some studies concentrate on women's home, domestic lives (see Davis 1988, Gerson 1985, Mackenzie and Rose 1983, Martin 1986, Massey and Wainwright 1985) whilst others concentrate on women's experiences within the workplace (see Deal 1988, Joseph 1983, Little et al. 1988, Mackenzie 1989, West 1982, Walby 1986).

Mackenzie (1989b) believes that it was women's daily activities, carried out in opposition to the contemporary urban form, and women's political actions in calling for day care, new designs for neighbourhoods, and better transit systems, which were instrumental in getting academics to notice that space was divided into men's public and productive spheres and women's private and reproductive spheres. The gendered literature therefore begun to see an increasing separation of the private domestic sphere and the public work sphere as they were explored individually. Researchers and employers often treated work and family as separate, independent systems, with the split between work (the subject of economic geography) as one sphere, and home (the subject of social geography) as another (Hanson and Pratt 1988a, 300). Bowlby et al. believe that during the 1970s and early 1980s, work on women and gender was treated as a specialised topic within geography, and it was felt that it could be ignored by those studying other areas of the subject rather than being integrated into all areas of the discipline. Academic studies were very insular and local culture or relations in the home were seen as important only in so far as they impinged upon or were affected by economic change (1989, 163). As Davis described, in reality there are two labour markets; one for men and one for women, with a very small overlap between them (1975, 98). According to Little et al. (1988), this second phase of feminist geography witnessed not only a social separation, but also a spatial separation between home and work, domestic and waged labour, public and private spheres and increasingly between the daily activity patterns of men and women. According to Campbell Clark (2000), this was due to the fact that men were assumed to have primary responsibility at work and women primary responsibility at home.

Dual Role

The recognition of the two separate areas of study brought with it a recognition of woman's dual role, for it became clear that she plays a part in both the private reproductive sphere and the public productive sphere. Freeman (1982) calls it the 'double shift' which is worked by many full-time employed mothers who have to complete the domestic role after getting home from work, as many women are attempting to fulfil their responsibilities for maintaining a home and community while at the same time performing public economic roles (Mackenzie, 1989b). It has been suggested that the inequality in employment patterns and opportunity can be linked to patriarchal relations in both home life and the workplace experience of women employees. Many socialist feminists (for example Walby 1986) have seen the patriarchal nuclear family as an important source of women's oppression; women's labour is exploited within the home and their domestic responsibilities dictate the terms of women's position in the labour market (Pratt 1993, 53). For as Zelinsky et al. point out, those women who did enter the labour market continued to carry out the bulk of the domestic tasks and home responsibilities, not shared by men regardless of their involvement in paid employment (1982, 348). Oakley (1981) believes that without this backup of domestic labour the economy would not function and asserts that women are the backbone of the economy. Mackenzie (1988) describes how women's allocation of time and adjustment of space must constantly attempt to balance the amount of time required from the mother, the wife, the cook, the cleaner and the wage earner, to achieve a desired family lifestyle. In addition to carrying out the demands of their multiple roles with less mobility than their male counterparts, Miller, after calculating time-budgets of women, concluded that women's free time was severely limited by their responsibility for the domestic work within the home. The difficulties that face women in combining their responsibilities restrict women's choice of jobs because they are frequently restricted to jobs which make the minimum demands or least conflicts with their home lives. This is regardless of these women's own desire for achieving challenging employment and career fulfilment (Women and Geography Study Group 1984, 70).

Separation of Spheres

Massey (1984) and Villeneuve and Rose (1988) both note that this spatial separation between the domestic, home sphere and the employment, work sphere is the key component in the study of the role of gender and class in the overall spatial division of labour. Although there was a growing separation within the academic literature between study of home life and paid employment, the Women and Geography Study Group noted that at this stage, geographers, in general, paid only cursory attention to the impact of the division. Geographers neglected to analyse both the impact of these changes on social relations between men and women, and the relationship between the organisation of domestic work and the urban spatial structure (1984, 43). The Women and Geography Study Group demonstrated how 'important such concepts of patriarchy and woman's dual role are, if geographers are to understand fully the decisions taken at various levels about the location of homes, workplaces, and public and private facilities *and* interrelations between these domestic decisions and the differing day to day behaviour, experiences and opportunities of women, men and children' (1984, 24).

What is important within this early phase is the way in which feminist geographers began to direct attention to the beginnings of a recognition of the link between home and work through academic research on women's dual role.

Hanson and Pratt argue that more individuals are trying to combine paid employment and home in a single daily activity pattern, rather than in two separate ones, meaning that the interdependencies between home and work need to be reassessed (1988, 303). In fact Granovetter and Tilly (1988) argue that understanding personal networks is central to understanding labour market inequalities, and if urban geographers are to be able to understand patterns of housing consumption, the meaning of work, home and neighbourhood, in fact the geography of cities, then the home-work links need to be re-examined and reconceptualised (Pratt and Hanson 1991, 303). Pratt and Hanson provide three ways in which the effects of the home on paid employment need to be incorporated into a reconceptualisation of geographic home:work interdependencies. Firstly the structure within the home can affect decisions about whether to work outside the home, hours of work and the type of work, therefore if geographers and policy makers wish to understand employment patterns, they must look beyond the workplace to the home. Secondly, what we mean by 'home' and 'work' needs to be redefined and thirdly, Pratt and Hanson suggest that the local context mediates the home-work link, for example the availability of social support and services in the local neighbourhood. The suggestion by Pratt and Hanson that the local is an important element in shaping the relationship between home and work, is a sophisticated development in relation to labour market theory at this time, and will be explored later in this book. The Women and Geography Study Group argue that the experience of women in the labour market cannot be understood without an analysis of gender divisions within the home, and indeed vice versa (1984, 145). In contradiction to those arguments, Walby (1986, 1990) and Reskin and Roos (1990) strongly believe that woman's position in the family is largely determined by their position in paid work rather than her position in the home. They argue that women's family roles are largely irrelevant to their occupational choices and that the origins of occupational segregation must be sought within the labour market. Walby argues that if our analysis is restricted to the current moment then it will look, superficially, as if the family significantly structures a woman's employment decisions. Walby believes that it does not provide an explanation of the structures which constrain a woman's choice (1990, 56). This argument was not continued until almost a decade later and I shall therefore return to it when I consider what has been labelled fourth generation segmentation theory.

Evaluation of the Dualist Model

First and second generation segmentation theory represented the development of the dualist model of labour market theory. There are two areas in particular in which the dual labour market model represented a significant break with what had gone before: firstly it shifted the emphasis away from the characteristics of jobs; and secondly it bought an understanding of institutional processes into mainstream labour market theory (Peck 1989, 122). However, there are also many criticisms of the dualist model. Firstly, Cooke (1983) describes that the dualist model unrealistically assumes that each section of the labour market is sealed off from the other with no integration – something that third generation segmentation theory explores further. Secondly,

Craig et al. (1985) and Peck (1996a) both argue that the dualist model concentrates on the characteristic of the jobs and not the workers, therefore assuming that labour supply factors play no direct role in shaping the pattern of employment organisation and of inequality created by the demand of labour. In relation to women workers, Peck (1996a) suggests that women are not innately suited to secondary sector employment by virtue of low attachment to the labour force. Although their dual role as domestic labourers and as wage earners is reflected in the tendency for women, as a group, to exhibit high turnover rates, it is also true that the inherent instability of secondary jobs is a determinant of high turnover. Although there are many criticisms of the dualist model, it helped, particularly in the hands of radical theorists, to open up a series of fundamental questions concerning the operation of labour markets (Peck 1996a, 56). Ashton and Maguire conclude that dual labour market theory was initially designed to further understanding of the operation of the national market, however, at the local level it was 'too simplistic' (1984, 118), which has led to the more multi-causal approach developed in third generation segmentation theory.

Through an exploration into the ways in which men and women differentiate, this second group of literatures identified woman's dual role; that of private domestic sphere and public work sphere. Researchers of feminist theory and labour market segmentation theory had previously treated these as separate independent systems, however recognition came that there may be a link between the work and domestic domain and that one sphere may have an impact on the other, and so research was needed into the way in which women combine their daily lives.

Linkage of Dual Roles

Third generation segmentation theory made the advance from the dual model to a more complex multi-causal approach through its acceptance of the interpenetration of the two spheres; public and private, and the influence and complexity of women's dual role.

Third Generation Segmentation

According to Peck (1996a, 57) third generation approaches represent a significant advance over the early dualist models, for it became less concerned with simple patterns and processes of labour market inequality, and increasingly began to explore the fundamental dynamics of the labour market. Peck believes that the labour market does not have a single institutional dynamic (1996, 77), but, as Rubery explains, it can best be approached via a cluster of models or theoretical approaches which have arisen out of labour market research during the 1970s and 1980s (1992, 246). By the third generation, labour market theorists began to appreciate that an explanation for participation within the labour market was not solely about what the labour market had to offer, but such participation was also impacted by the process of social reproduction, including factors stemming from the domestic division of labour. Peck argues that segmentation is partly a product of who does the jobs, social conventions concerning appropriate forms of waged and unwaged work undertaken by different social groups, and the relationship between the labour market and the sphere of social reproduction (Peck 1996, 72). This led to an understanding that economic, social and political forces combine to determine how economies develop (Wilkinson 1983, 413), and so multi-causal explanations have increasingly been sought by third generation theorists, a recognition which was absent from dualist models.

Third generation segmentation theory also brought with it an understanding that a full appreciation of women's position in the labour market could only be gained if it was acknowledged that the two spheres, those of home and work, interpenetrate, each conditioning each other. According to Peck, a notable feature in recent segmentation research, representing a considerable advance on the dualist models, has been the specification of the role of labour supply factors in structuring the labour market. Peck believes that three areas merit detailed consideration: social reproduction and the role of the family; labour union structures and strategies; and the position of marginal groups in the labour market (1996, 65-6). Bauder (2001) explains how the notion of the 'social nature of labour' (Peck 1996, 29) recognises that workers are social actors as much as they are labourers, and that social division is constructed outside the market and then shapes employment relationships.

Similarly to labour market theorists, the geographers Pratt and Hanson argue that the role of the family household is an absolutely central one for those studying economic processes (1991, 56). According to Craig et al. (1985) the family exerts an important influence on labour market activity in three ways: first, it plays a key role in the social conditioning and education of the young; second, it provides support for workers in the labour market as well as for other dependents (such as the sick and the old); and third, the sharing of income within the family unit impinges on both male and female roles in the labour market. Employers' expectations are important here as males are expected to be able to extract a 'family wage' from the labour market, while women are perceived to be in search of no more than 'pin money' (Peck 1989, 130). Workers' expectations were conditioned both by their perception of employment opportunities and by their position in the family, which determined their income needs and domestic responsibilities. Segmentation theorists became aware that the institution of the family serves to structure the supply side of the labour market in the allocation of domestic responsibilities. Pratt and Hanson (1991) explain that women still bear the burden of domestic responsibilities they believe that this is evidence that women are exploited and oppressed by men in the private household. Divisions of labour, both in the labour market and within the home, are constructed around the notions of male 'breadwinner' and wage-earning roles and female domestic labour and child-rearing roles. As Marshall explained, there is now a 'recognition that men and women supply their labour on different terms' (1994, 44), reflecting role specialisation within the family and the complexity of social expectation attached to roles.

   This third set of literatures highlights the development of a multi-causal approach to both labour market policy and the way in which women negotiate their daily lives. It was acknowledged that a full appreciation of women's labour market participation would be gained by acknowledging that the role of the family household is central to the process.

The Differentiation of Women

Feminist literature recognised the need to deconstruct the identity of woman after socialists such as bell hooks (1984) suggested that different women experience systems of patriarchy in different ways dependent upon race, religion and life stage. Fourth generation segmentation theory acknowledged the need to consider a spatial dimension to the labour market, highlighting that labour markets operate in different ways in different places.   Therefore women's experience of paid employment will vary accordingly to the operation of the locally specific sets of relations for example, the promotion of a certain lifestyle, either that of working or not working, and the provision of local services such as public transport and childcare. Academics in both sets of literature began to question whether women actually had a choice to work, not work or combine the two, and began to look at what factors influenced that 'choice', such as employer attitudes, family-friendly policies, childcare issues or the existence of patriarchal systems of practice.

The Deconstruction of Woman

Whilst segmentation theory was developing the connection between home and work and the role of the family, feminist geographers between the late 1980s and mid 1990s were deconstructing the category of 'woman'. Feminist geographers were increasingly attentive to the differences in the construction of gender relations across races, ethnicities, religions, sexual orientations and nationalities. This third phase of feminist geography is characterised by the recognition of difference, not only between men and women, but also differences which exist between groups of women. Up until this point, most of the research has dealt with urban women in western societies, but now there was a recognition for a need to examine rural women, women of different ethnic and racial groups and women at different points in their life cycle, making some cross-cultural and social comparisons (Zelinsky et al. 1982, 352).   Rose (1995) argues that the universal claims of feminist geographers in fact excludes some aspects of women's experiences, and according to the Women and Geography Study Group gender roles and the lives of women vary quite dramatically even within so small and comparatively homogeneous a country as the UK (1984, 12). For labour market theorists, this recognition that women were not a homogeneous group, led to an interest in the variation in women's experiences of labour market interaction, which was developed later in phase four.

Fourth Generation Segmentation Theory

A long list of factors which influence labour market segmentation does not constitute a theory of the labour market. Segmentation research must, therefore, explore in greater detail the inter-relationships between labour demand and labour supply, both of which are dependent on the local labour market. Peck therefore, believes that comparative research would be particularly valuable in this regard, as international comparisons, or perhaps comparisons between contrasting local labour markets (1989a, 138). As Kreckel (1980) explains, modern labour market theory has become more open to sociological considerations it is no longer concerned solely with patterns and processes of labour market inequality, but is increasingly moving to explore the fundamental dynamics and social foundations of the labour market (Peck 1996a, 74. See also Bauder 2001). In this view, the supply of labour is not governed simply by market forces, but also by demographic factors and social norms concerning the participation of different groups in wage-labour, which is mediated through the institution of the family (Peck 1996a, 75). Peck acknowledges that labour market participation is profoundly shaped by the networks of spatial and temporal routines which women must construct between waged work on the one hand, and various and caring tasks on the other (p334). Segmentation theorists have begun to realise that the family is a main factor in influencing a woman's decision to enter the labour market through the negotiation of women's dual role, particularly the domestic division of labour, and that to deal with the spatial dimension – to come to terms with geography – represents an important stage in the development of segmentation theory, giving us the basis for a *fourth generation* (Peck 1996a, 86). Peck believes that

> '"work" and "home" are not the hermetically sealed, separate spheres they were once assumed to be, but interpenetrate in a variety of ways, strongly conditioning the job-market behaviour of both men and women.'
>
> (2000, 142)

This argument has also been recognised in feminist geographic thought by Hanson and Pratt.

> 'Not only have relatively few of those who have written on dual or segmented labour markets considered the spatial aspects of such segmentation, but those who have recognised the spatial dimensions of labour market segmentation have tended to conceptualise such dimensions as existing at a scale no smaller than that of a region or the metropolitan area as a whole.'
>
> (Hanson and Pratt 1988b, 181)

Fourth generation segmentation theory thus acknowledges a need to explore the structures and dynamics of local labour supply and recognises that labour market segmentation is, at least, partly a locally constituted process. In this respect, labour market theorists have become aware that it is a combination of the geographical unevenness in childcare, social moves regarding gendered distribution of waged work, the local labour market processes and workforce politics, welfare entitlements, male and female participation rates and levels of single parenting which affect the local constitution of the secondary workforce (Peck 1996b). Hanson and Pratt explore these issues and analyse how the results of various studies documenting women's commuting distances to paid employment, show that the commuting ranges of many women tend to be smaller than those of most men, creating a number of separate labour markets within any single large metropolitan area. This suggests that the geography of gender relations is constructed at a very local scale and that the experiences of gender and of occupational segregation may be different from place to place even within one metropolitan area. Hanson and Pratt (1995) have highlighted how home-work interdependencies are the core to an explanation of the local labour market. They state 'that the home-work link which is already embedded in definitions of the local labour market should be expanded and placed at the analytical core of local labour market studies' (1991, 375). Likewise, Odland and Ellis believe that 'women's labour force experience may vary across metropolitan areas, even for women of similar personal characteristics, because women encounter different conditions in different places' (1998, 334), and therefore, as Granovetter and Tilly (1988) argue, understanding personal networks is central to understanding labour market inequalities. Bauder (2001) argues that the task for labour market theorists is to gain a further understanding of the complex and interlocking relationships between the labour demand and supply-side processes. In order to do this, Bauder believes that fourth generation segmentation theorists should focus on place as a constitutive force therefore demanding further research at the local level.

Down to the Local

Massey argued that the construction of gender relations is also strongly implicated in the debate over the conceptualisation of place (1994, 7), and in parallel to segmentation theory, geographers also began to recognise the importance of the local. Hanson and Pratt (1988b) point out that to date, labour market research has tended to focus on national trends and national variation, with little appreciation given to local geographies of participation and progress by women which, according to Odland and Ellis, conceal 'important local and regional differences in women's labour force activities' (1998, 333). Odland and Ellis discuss how most research on local labour markets has concentrated on highlighting place-to-place variations in wages and unemployment, and only a few studies have questioned whether participation in the labour market depends on 'the special character of the local or regional labour market' (1998, 336). Hanson and Pratt (1995) argue that state policies and labour supply and demand processes come together in different ways in different places to create a locally variable pattern of labour market segmentation. One example of this came from Mattingly, Hanson and Pratt, who describe how data from their Worcester, USA study showed that women's career decisions during and after child-related breaks were strongly influenced by the local context in which they lived (1998, 11). Bauder (2001) recognises that workers are social actors as much as they are labour, and the social division constructed outside the market, shapes employment relationships. Odland and Ellis also believe that 'women's participation in market work is strongly associated with their individual characteristics...and a metropolitan area may differ in ... the frequencies of these characteristics in their female population.' They explain that if an area is made up of a variety of characteristics and life stages 'then localised differences in demography, rather than localised differences in labour market conditions, are the primary determinant of the geography of women's labour force experience' (1998, 334). After acknowledging such differences in labour market variation, geographers such as Hanson and Pratt have begun to explore these issues, saying that 'factors such as the local availability of support services – whether provided through informal social networks or through more formal public or private sources – can enable or constrain people's ability to work outside the home' (1988a, 308). Tivers (1985), in her research with women who have children under five years old in south London, discovered that women did want to enter paid employment, but were often prevented from doing so by the lack of local services such as public transport and childcare facilities. Almost all of the women that Tivers spoke to who were in paid employment, worked part-time and locally in low-status jobs. Hanson and Pratt argue that 'it is undoubtedly the case that the neighbourhood is an important 'agent' of socialisation that structures work/career aspirations and work related attitudes' (1988a, 309). It is through geographers' work on the deconstruction of the identity of woman, and the importance of spatial variation in labour market theory, which has allowed researchers such as Hanson and Pratt to explore and fully understand the impact of the local environment on women's lives and their decision regarding their 'choice' to enter paid employment.

> 'When we first began to write about women and work in Worcester, we stressed the homogeneity of women's work experiences...We laid the sameness of women's work experiences against a far richer residential geography, one patterned by difference of family type and occupational standing (Pratt and Hanson 1988). We are now attuned to a more complex layering of geographies and see a multitude of differences across communities located within the same metropolitan area.... These differences create different frameworks within which families construct their lives and experience gender.'
>
> (Hanson and Pratt 1995, 185-6)

Massey questions if this cultural link between women and locality has influenced reasons for women's issues being ignored for so long. 'Is it *only* a local struggle, *only* a local concern?' (1994,10). However, later in her book when Massey presents a nineteenth-century case study of coalminers' wives, she demonstrates, as many others have done, how 'the local has its impact on the operation of the national' (1994, 192). This asserts that geographers need to study and understand women *and* the local in order to gain a full understanding of national economic production. Hanson and Pratt believe it is the twin processes of employer intentionality (i.e. locating firms so as to capitalise on certain types of female labour) and women's spatial entrapment (i.e. their inability or reluctance to travel long distances or times into work and their subsequent reliance on limited local employment opportunities) that are certainly the most frequently postulated reasons for expecting spatial segmentation within the labour market (1988b, 184). Bauder (2001) believes that the notions of the local labour market must relate to place-based contexts that define cultural identity and signify cultural difference. According to Bauder, the crux of spatial segmentation theory is that local labour markets are 'conjunctional structures' that operate in different ways, at different times, in different places' (Peck 1996a, 94).

Contemporary Work on Geography and Gender

Contemporary direction on geography and gender has developed from the early 1990s and is characterised by firstly, a greater recognition that each sphere – home and work, public and private, men and women – cannot be treated separately as they have previously been, for each realm has an influence upon the other whether directly or indirectly. McDowell believes that recent work has led to an insistence on diversity in the definition and analysis of gender relations, challenging the over-simple distinction between man and woman, as social constructs and categories for analysis (2000, 512). An example of the overlapping of the two spheres, that of public and private, is described by Rubery (1994) as she explains that it is not possible to look at the high proportion of women who work part-time in the UK without analysing the full set of reinforcing factors that have led to this outcome. These reinforcing factors include favourable social security systems, the dominance of large firms in the service sector which have developed sophisticated working-time planning systems, the lack of childcare facilities and the establishment of standards of living based on a norm of a male breadwinner on long hours and a female part-time worker. Likewise, Walby (1997) describes how our view of the system of gender relations in this phase was changing, from one which is based on women being largely confined to the domestic sphere, to one in which women are present in the public sphere, necessitating an examination of the interaction between public and private. Secondly, we have witnessed a greater recognition of the spatial variation in factors that influence women's lives, highlighting local differentiation. For instance McDowell has explored the very specific sets of gender relations which have developed in the City of London (McDowell 1997). Her work explored gender relations in one of the most rapidly expanding sectors of the economy: financial services, where she looked at the ways in which everyday practices in merchant banks reinforce gendered roles.

Despite the acknowledgement that women are playing an increasingly vital role within the new economy, reflected in their increasing participation in the formal labour market, and despite contemporary government policies to encourage women into the workplace such as the New Deal initiative, women still remain the primary carers for their children and maintain the responsibility for organising childcare to facilitate their entry into the workforce. McKie et al. (2001) believe that even in the twenty-first century, employers still tend to believe all women are less reliable workers due to their potential commitments outside of their employment. Due to the acceptance of the impact of the home on women's participation in the formal labour market, academics have argued for further research into the formation of gender roles within the home and the relationship between home and work (see McKie et al. 2001, Gerson 2002, Marks et al. 2001). According to Benn, the public concern of the private world of caring has raised the question of how to reconcile work with parenting (1999, 232). Although academics see the need for such research, McKie et al. believe that in relation to policy,

> 'The main site in which gender roles are created and reformed, the home, is excluded from any external examinations or debates in current consultations.'
>
> (McKie et al. 2001, 240)

This book allows the home to be placed at the forefront in order to discover the extent to which it influences women's participation in the formal labour market. In attempting to understand women's decision-making process and issues of home and work, contemporary debates have highlighted the fact that women are a heterogeneous group (see Hakim 1993, WU 2000b). It is acknowledged that some women may wish to combine paid work with parenting and other women make the positive choice to remain at home and be the full-time carer for their child(ren) (see McDowell 2000, Deseran et al. 1993, Odland and Ellis 1998). This is further explored by Hakim through her use of Preference Theory.

Preference Theory

Hakim argues that social science research has tended to treat women as a homogeneous group, one which naturally seeks to combine employment and family work (2000, 6). However, Hakim argues that women are a heterogeneous group especially in terms of the preferences and priorities they express and make concerning the conflict between family and employment. It has recently been claimed by Hakim (1995, 2000) that women's relatively disadvantaged position in the labour market is not a consequence of the institutional and/or structured disadvantages they suffered. Rather, women's position reflects the outcome of their varying choices (Crompton and Harris 1998, 118). Hakim (1995) argues that there are five feminist myths about women's employment;

1. The myth of rising female employment.
2. The myth that women's work commitment is the same as men's is often adduced to resist labour market discrimination.
3. The myth of childcare problems being the main barrier to women's employment.
4. The myth of poor quality part-time work is used to blame employers for the characteristic behaviour of part-time workers, including a high labour turnover.
5. The myth of exploited part-time workers.

Hakim argues that the reason the above are feminist myths is that they are developed on the assumption that women reject the role of full-time homemaker, and seek to participate in the labour market on exactly the same basis as men. Hakim believes that it is a myth to think that as soon as these barriers are removed women will flood into waged work on a full-time basis (1996c, 179). In Hakim's Preference Theory she asserts that approximately 40 per cent of women have made a positive choice about their involvement with children; 20 per cent are home-centred, who have made a choice to be at home with their children full-time, whilst the other 20 per cent are work-centred and have either made the decision not to have children, or use a crèche, have a nanny or a househusband. This is endorsed by Benn (1999) who believes that having children in the twenty-first century has become a choice rather than an expectation. Within Hakim's Preference Theory, this leaves approximately 60 per cent of women who are adaptive, these are the women who are trying to combine home and work. Hakim asserts that most women have therefore made the choice to live with a dual role, a belief which is supported by Ginn et al. who argue that it is possible to be highly conscious of the needs of one's family and at the same time to care deeply about maintaining employment (1996, 168).

Hakim believes that, until recently, research on women's work and gender issues has been on the sex-role prescriptives offered to women and on the situational constraints on their behaviour – on what they are *expected* to do and what they are *prevented* from doing, but never on what they *want* to do (2000, 14). According to Hakim, Preference Theory does two things: it reinstates (heterogeneous) personal preferences as an important determinant of women's behaviour, and it states that attitudes, values and preferences are becoming increasingly important in the lifestyle choices of people in rich modern societies (2000, 17). Hakim's Preference Theory argues that within the twenty-first century women do have a choice, and that feminists should acknowledge, through the literature on the deconstruction of woman, that women are not a homogenous group. Hakim argues that a woman who has given up a professional career to be at home with her children full-time may well have made a positive choice, and it should therefore not be assumed to be a position into which she has been forced.

Choice to be at Home

Some women make a positive choice to mother full-time (Holloway 1999). Hakim describes the home-centred woman as one who accepts the sexual division of labour in the home, prefers not to work, and to give priority to children and family life throughout their life (2000, 159). Once married, home-centred women prefer to be full-time homemakers, and childrearing activities are of central importance to them. It is possible that home-centred women may return to work after caring responsibilities lessen, but only if the employment fits completely with their home-centred lifestyle (2000, 159).

Choice to Work

Increasing numbers of women return to work between births, take shorter spells out of the labour market between births and return to work more quickly after completing their families (Martin, 1986). The presences of the 'gender role constraint' as described by Tivers (1977) provided the social expectation that women's main activities should be those of family care, and those women who do 'choose' to go out to work often face discrimination from others. Although a lot of this discrimination comes from men, as Walby (1997) and Holloway (1999) explain, this is not always the case, it also comes from friends and relatives, for example women of an older generation who may be jealous of the opportunities the younger generation of today have. Different family generations have come to have different personal expectations due to the societal aspirations they have grown up in. Today's younger generation believe that they should have a genuine choice to go out to work, however many of the older generation believe that a mothers' place is at home with their children, not in the workplace while someone else looks after them. There have been discussions on downward occupational mobility upon returning to work after a break to have children, however, Mattingly, Hanson and Pratt argue that whilst many women do return to jobs of a lower status than the ones they left, the change is not necessarily an effect of the break. If a woman's break involved adding a child to the household, she has substantially more domestic responsibilities upon returning to work than when she left, which can influence her job choice (Mattingly, Hanson and Pratt 1998, 17). The predetermined societal expectation can also have an affect upon men who become the 'househusband' and main carer for the children whilst the woman of the house becomes the main breadwinner. Members of society will often quiz a man as to how he got into the situation, and perhaps take pity upon him for it is still not an expected situation within western society.

Balance

The contemporary woman, who goes out to work as well as looking after her family, needs to find a balance between paid employment and the home. Can there ever be a true balance? Daily Telegraph journalist Anne McElvoy coined a new term for a new kind of the woman; 'Hyperwoman has a family, a high-powered job and a house to run. A woman with everything but time' (Benn, 1999, 104).

Regardless of a woman's employment status, she often feels that it is her responsibility to run the family home. The attitude of men enhances this feeling of responsibility partly through the idea that men's help in the home, is just that, help (Benn 1999). Luxton, in her research into the gendered division of labour in the home, discovered that men are prepared to help out, 'for example, a number of men did the grocery shopping on a regular basis but they insisted that the women draw up the basic list of things needed. Similarly, some men would do the laundry, if the dirty clothes were previously collected and sorted and if the necessary soap and bleach were at hand' (1990, 48). Should it not be their equal responsibility rather than an act to help the woman?

According to Hakim (1995b, 1995c, 1996), you can never have an equal balance between home and work. Hakim believes that women are either 'Careerists' whose commitment to professional advancement takes precedence in their lives or 'Home Centred', for whom paid work is a subsidiary activity, which does not take priority over domestic responsibilities paid work is simply an extension of their role in the family. According to Dex et al. (1995), and later Hakim (2000), most women want to enjoy both work and home life to the full, however, there can be a resentment by women trying to live in two worlds, that of mother and career woman but, in fact, never being fully present in either (Benn 1999, 78). However, the weighting of the balance between paid work and domestic labour is not always a choice and is often influenced by the attitude and the amount of responsibility the partner takes within the household. Benn believes that 'new motherhood can only take the jump from image to reality if a real new fatherhood becomes a reality' (Benn 1999, 246). In order to achieve this, Pahl (1985) suggests that household work strategies have to be adopted. This relies on both members of the household accepting equal responsibility and agreeing to adopt 'distinctive practices' in order to complete household work. Besides the partner, childcare is one of the most widely discussed barriers that can prevent women entering or not entering into the labour market (see Dex 1986, DTI 2000, Holloway 1999).

Childcare Provision

The provision of childcare has become an important issue in labour market studies. Access to childcare bears significantly upon women's employment choices and decisions (Tivers 1977, Palm and Pred 1974), for inadequate childcare often forces women to withdraw from the workforce. Zelinsky et al. (1982) describe how, without appropriate childcare services, women are likely to be confined to part-time jobs or jobs close to home which may be less rewarding than other work, and below the status merited by their qualifications and experience. Childcare is often described as being the main factor which dictates women's participation in paid employment: however, as Hakim describes through her Preference Theory, home-centred women do not need childcare services and work-centred women do not hesitate to pay the cost of childcare services. It is adaptive women, who are torn between the desire to work and the desire to be a full-time homemaker, who are most likely to demand affordable high quality childcare services provided by the state, in order to improve their flexibility of choice and ease their guilt (2000, 176). Hakim is keen to point out that this is one group of women and should not therefore be promoted as the only issue, common for all women.

Part-time Work

One way in which women find a satisfactory compromise between their wish to look after their children and their desire to work is through part-time employment. This offers women the opportunity to still retain an income and benefit from other positive aspects of going out to work such as social interaction and earning their own money, yet at the same time they are often still able to take the children to school and/or pick them up. However, as Hanson and Pratt (1995) point out, women who choose a part-time schedule as a strategy to combine paid employment with mothering work, move within particular constrained spatial orbits, either financial, geographical or emotional. According to Hakim, the vast majority of part-time workers are married women, and in many respects, women working part-time more closely resemble wives who devote themselves full-time to their responsibilities than to women in full-time employment (1993, 104-5). Many mothers find working part-time a necessity or an attractive arrangement because finding adequate childcare is difficult. Hakim (1995) argues that the dominant feminist view continues to insist that part-time work is an unwilling choice forced upon women by the need to cope with childcare responsibilities, a compromise rather than a preference. Hakim argues that this view fails to take into account that part-time workers have the highest levels of job satisfaction despite being restricted to the least desirable jobs, nor the fact that childcare problems do not prevent large numbers of mothers from working full-time. As Watson and Fothergill point out after their research into attitudes to part-time work, the key problem with this argument is that the popularity of part-time work, and of not working at all, extends well beyond women with childcare responsibilities (1993, 214).

A Critique of Hakim

Female labour force participation is not simply about negotiating a choice within the household, finding childcare provision or about women claiming the right to go back to work, it is a more complex process that combines all three. Crompton and Harris (1998), whilst acknowledging that both men and women make choices, argue that their choices are constrained by different factors including work-life biographies and life stage. Their work, explaining women's employment patterns, found that women's biographies demonstrated that orientations to employment and family life were complex and variable, and although Crompton and Harris accept that preferences may shape choices, they argue, contrary to Hakim's assertions, that they do not determine them (1998, 131).

Similarly, although I acknowledge Hakim's assertion that many women have made a positive choice to be in their current situation, be it either to be at home full-time, to not have children at all or to combine their working careers with bringing up children, I feel that Hakim's work does not explore the geographies of preference. In order to provide a fuller understanding of women's labour market participation, spatial aspects need to be explored and added to the Preference Theory. By this, I am referring to the spatial aspect which has been developed and accepted by fourth generation segmentation theory and in academic debates on locality (see Duncan and Savage 1991, discussed in more detail in chapter 8). Furthermore, the acknowledgement that labour markets operate in different places in different ways (Peck 1996a, xv) needs to be integrated into Hakim's Preference Theory in order for it to provide a further understanding of women's labour market participation.

Genuine Choice?

The notion of 'choice' cannot be accepted without question. The first hurdle for a woman is openly expressing her lifestyle preferences. Hakim (2000) believes that preferences which are contrary to social norms can be concealed or left vague, so as to avoid argument or opprobrium. As an example Hakim argues that the desire to have children is rarely stated explicitly because it is natural, obvious and unquestioned. However, the desire not to have children may also be left unstated because it seems to invite an argument in ordinary settings (2000, 16). Can a woman go out to work without facing discrimination from others who think she should be at home with her children? Can a woman choose to stay at home for a few years to bring up her children, without penalising her future employability? According to Benn, compared to even ten or fifteen years ago, the burden of explanation and justification is now on the woman who doesn't have a job (1999, 33). Is the decision to work a genuine choice or a financial necessity? Does the woman go out to work due to the pressure from the partner or society? Does the promotion of female dominated labour apply added pressure to women to go out to work? Benn asserts that the media has contributed to the pressure on a woman to become carer, mother and worker.

> 'Soap operas on radio and television rarely feature stay-at-home mothers, especially among the younger generation.'
>
> (Benn 1999, 35)

Walby, however, believes that the decision to spend more time on the home or more time on paid employment, particularly choosing female-dominated employment, is a rational choice (Walby 1997). Similarly, Weedon presents that women's perceived oppression in the labour market, may well be one of genuine choice.

> 'If as a woman I know that I am naturally unaggressive, emotional, unself-confident, unambitious, caring, tactful and have a strong need for security, and that this means in social terms that my primary location and sphere of influence is in the patriarchal nuclear family where I spend my life on housework and childcare, does that mean that I am oppressed? If I make a conscious decision to earn my living by selling my body for prostitution or pornography, am I oppressed? Surely if a woman's choice is based on her free will, then it must be valid.'
>
> (Weedon 1987, 84)

Conclusion

Since its initial impetus, feminist thought has changed its focus and direction several times, from the universality of woman to the social differences between women, from women being close to nature to femininity as a social construction, from modern to post-modern, and from structuralist to poststructuralist. Whichever direction of geography has been studied, one common belief is that feminist geographies trace the interconnections between all aspects of daily life, across the sub-disciplinary boundaries of economic, social and cultural and therefore cannot be ignored (Pratt 1994). Feminist geographers have noted that those women who entered the workplace tended to still be responsible for the majority of domestic tasks, and consequently had strict restrictions set against the type and amount of paid work they could undertake. However, it is only recently that geographers have tried to discover 'why' this is so, and believe the understanding comes from a detailed study of the interdependencies between home and the workplace.

It is Hanson and Pratt (1988a) who first brought to my attention the concept of the decision to work as a social process formed through intra-household processes, and they argue that if geographers want to understand employment patterns, they must look beyond the work place to the home. Labour market segmentation theory has also come to acknowledge the importance of the home in the process and their argument runs parallel with Hanson and Pratt who believe that aspects of men's and women's employment patterns can be illuminated by studying intra-household social relations, identifying ways in which the effects of the home on paid employment need to be incorporated into the reconceptualisation of geographic home-work interdependencies. Although phase four in the development of the literature represents a new way of thinking and the recognition of interdependencies, this is not a new concept, it has simply become more explicit in contemporary feminist thought. In early feminist thought the Women and Geography Study Group acknowledge that 'although in this introductory book we decided to look in turn at cities, work, access to facilities and the development process, we have tried to show you where the links between these areas of research exist. We have argued that the experience of women in the labour market can not be understood without an analysis of gender divisions within the home, and indeed, vice versa' (1984, 145). Simultaneously, labour market segmentation has also acknowledged the need to look beyond the labour market into the home for an explanation of participation patterns and suggests that local economic and social processes may vary in accordance with locality. However,

> 'If the local labour market is to be raised from the lowly status of a case study area to that of mid-level theory, the pressing need is for the theoretically informed, concrete research at the local level. While there are numerous suggestive studies (for example, concerning the operation of rural labour markets, the link between labour markers and local

politics, the construction of gender divisions in local labour markets, and urbanisation and the labour market), theoretically informed, local labour market research remains in its infancy.'

(Peck 1996, 112)

My research addresses Peck's agenda by undertaking theoretically informed local labour market research into the geographies of women's labour market participation.

Chapter 2 has sought to highlight links between two sets of academic literatures; labour market segmentation theory and the development of gender studies within geography. It has also sought to identify how we can take elements from both in an attempt to move forward in understanding women's labour marker participation patterns in the UK. In the next chapter I shall examine what women's position is within the UK labour market and explore the current demands women have in order to be able to fulfil that position.

Chapter 3

Women and the Labour Market: The Shifting National Context

Introduction

In the past fifty years there have been dramatic changes in the attitudes of, and perceptions towards, women, which have brought with them a shift in women's opportunities and expectations regarding work. This chapter outlines women's position in the UK labour market from a national political perspective and provides a context for the detailed empirical work which follows later in the book. It demonstrates how, and suggests reasons why, women's labour market participation has dramatically increased over the past fifty years, followed by a closer examination of today's 'knowledge-based economy' and why women are a crucial component to its success. After an examination of contemporary government initiatives in Britain and policies affecting women, this chapter will demonstrate the limitations of national political strategy and highlight the mixed messages that the Government are sending to women in their attempt to promote 'choices'.

The Government promotes its aim to provide women with *genuine choices* in order to enable them to maintain a work:life balance through its recently launched Work Life Balance Campaign and the National Childcare Strategy. I shall argue that the proposed modifications in legislation are the absolute minimum changes necessary to enable women to think about options with regard to their employment; however, the Government does not go far enough in recognising how individual women will manage to fulfil their role in the new economy satisfactorily.

Historical Perspectives on Labour Market Participation by Women

One of the most dramatic changes which has occurred for women in the past fifty years is particularly evident in the sphere of paid formal employment, where there are now more women in the formal labour market in the UK than ever before. By the start of the new millennium, 75 per cent of the total UK population of working age was in employment, which is the highest participation rate in Western Europe and higher than in the USA (WEU 2001, 1). As can be seen from figure 3.1, the UK has the fourth highest female employment rate in the EU behind Denmark, Sweden and Finland, and is above the EU average in every UK region.

Figure 3.1   Employment rate for EU member states (2000)

EU member state	% Women
Austria	59.7
Belgium	51.9
Denmark	72.1
Finland	65.2
France	54.8
Germany	57.8
Greece	41.3
Ireland	53.2
Italy	39.3
Luxembourg	50.0
Netherlands	63.4
Portugal	60.4
Spain	40.3
Sweden	69.7
United Kingdom	64.5
EU average	53.8

Source: Eurostat, Labour Force Survey

For women, the increase in employment rate has been dramatic in comparison to men. The number of economically active females increased by 45 per cent between 1931 and 1970. In the same period, the number of married women going out to work increased fourfold. The number of male workers, meanwhile, remained stable (Coote and Campbell 1982, 49). In spring 2000, the number of women in employment had risen by 843,000 since 1990, while the number of men in employment was only 33,000 higher (Bower 2001, 107). Consequently there are currently 70 per cent of women of working age employed compared to 80 per cent of men (McDowell 2001a, 449).[1] As figure 3.2 demonstrates, these recent changes further close the gap between male and female participation rates, which has been steadily narrowing over the past thirty years.

---

[1] It will become evident through this chapter, there are a variety of statistics detailing women's participation within the labour market which do not always match. This is due to a combination of a variety of sources providing the information and the Governments generosity in their interpretation.

Figure 3.2   Proportion of men and women economically active in UK

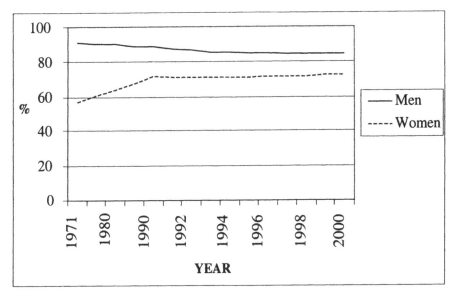

Source: Calculated from *Labour Market Trends* (August 2000 Table A.1)

When looking at the changes in labour market participation more closely, there is a particularly noticeable rise in the proportion of mothers with children under five who are in paid work, which has increased from 28 per cent in 1980 to 53 per cent in 1999 (WEU 2000a, 3). Additionally, McDowell highlights the increase of labour market participation by women in their thirties with dependent children (2001a, 449), as trends have shown a delay in women having children as they allow time to adopt and progress in a career before having a family. The proportion of 'traditional' households comprising of a couple with at least one dependent child fell from 38 per cent in 1961 to 23 per cent in 1998 (EOC 2000a, 2). Coote and Campbell argue that this increase in women's participation in the workforce represents a significant change for women. They believe that gone are the days when marriage, or the arrival of the first child, was expected to lift women out of the labour market and deposit them at home for good (1982, 49). One of the things this book will explore is whether the past expectations of woman as homemaker have really gone, or whether they are still prevalent in the twenty-first century.

These changes in women's labour market participation have arisen as a response to a range of factors including the rise of the women's movement in the 1960s, changes in reliable contraception, new patterns of consumption, and the restructuring of the labour market through the introduction and implementation of new technologies. More recently, McDowell argues that in its first four years, New Labour formed a government that combined the rhetoric of social democracy with neo-liberal policies, especially in the welfare sphere, which encouraged women to enter the labour market (2001, 448).   Giddens refers to this as 'The Third Way', in the sense that it is an attempt to transcend both old-style social democracy and neo-liberalism (1998, 26).

The 'new knowledge-based economy' which we are experiencing relies more heavily on female labour than ever before. Leadbeater (1998) defines the five principles of our new knowledge-based economy, which demand the inclusion of all, if it is to succeed. Firstly we must invest heavily in the creation and spread of knowledge. 'Knowledge has become the dynamo of economic growth through a powerful combination of complementary factors...These days the only thing that holds us back is our inability to make the most of our combined brain power...The ability to generate, apply and exploit our combined know-how is central to competitiveness in the modern economy' (p235). Secondly we must innovate and include everyone.   In a knowledge-driven economy, economic growth and social inclusion must go hand in hand. Thirdly, our goal must not be purely economic for we will only create a successful knowledge economy if we create a society that takes a democratic approach to the spread of knowledge.   Fourthly, a knowledge economy creates a global agenda which Leadbeater argues puts people in a position to make better-informed choices about how to care for their children. Finally, there is no best route to a knowledge economy. The old economy was dominated by the search for 'the best way' to do things, yet in the knowledge economy, driven by creativity and imagination, there is no best way (p239): these are skills which are more suited to the female workforce and so employment within this group has risen.

These increases in female participation look set to continue. As the UK moves away from a manufacturing industry towards a service-led economy, the Government projects that by 2011 there will be 1.7 million new jobs, of which 1.3 million are forecast to be filled by women (WU 2000a, 1). The jobs will be created through the continued expansion of the service sector where the types of jobs which are created are often part-time and shift-work. These are convenient and pragmatic for women as they can often be negotiated around a woman's domestic role; for example working 10am to 3pm enables a mother to drop her children off at school, go to work, and return to pick the children up at the end of the day. Alternatively some women like to work outside standard hours when their partner can be at home to look after the child(ren). Currently, in the UK, 88 per cent of jobs taken by women are in service industries, particularly in education and health (Bower 2001, 116). This pattern is common in every member state of the EU where more service jobs are organised on a part-time basis than any other type of occupation and the highest overrepresentation of female part-timers are found in this area of employment (Rubery and Fagan 1995). Women's share of managerial and administrative employment has risen since 1981 and it is projected that they will comprise 41 per cent of employees in this occupational group by 2006, compared with 23.5 per cent in 1981. The rise in the female proportion of employment in professional occupations is also set to continue, and by 2006, women are projected to make up 42 per cent of employment in this occupational group (Wilson, 1999, 13). It is projected that 57 per cent of women aged 16 and over will be economically active in 2011 (EOC 2000b, 4). As we shall see later, there is some confusion over the messages which the Government sends out regarding women and work, however, one point which is clear is that 'women's importance in the job market is growing and it is not overstating the case to say that the future success of the UK economy depends on women being able to reach their full potential' (WU 2000b, 6).

Unpacking the New Economy

The Government believes that;

> 'In the new economy, women will take time out of the workplace to have and raise children. But they will want to return to work because their jobs are fulfilling and businesses will understand that losing them represents a huge loss of skills, quite apart from the cost of replacing and retraining someone new. In the new economy, the old distinction between full-time and part-time work will blur as employers introduce a whole range of new working patterns to suit both their needs and the needs of their employees.[2]'
>
> (Jay 2001a, 6)

The Department for Employment and Education (DfEE) (2000a) stated that as we move into the new century, skills and learning must become the key determinants of economic prosperity and social cohesion of our country, and our entrance into the 'new economy'. McDowell refers to this as the 'new flexible post-Fordist economy' (2001a, 450), in which the key components are technology, knowledge, entrepreneurism, and the skills and experience of individuals. Moreover this move to a new economy 'will only be successful if it utilises the skills and talents of all women' (Jay 2000a, 2). The Government says that,

> 'in this new "knowledge-based" economy, our future prospects will depend on the ability to liberate the human potential of every citizen. It means that as brain replaces brawn as the driving force at work, women's skills and contribution will become ever more crucial to the success and improved productivity of our economies. And the skills that women excel in – such as project management, interpersonal skills, high standards of customer service – are all highly valued in this changing economy.'
>
> (Jay 2001b, 3)

---

[2] For examples of new working patterns see figure 3.4.

The Government acknowledges that women bring different skills and strengths, to the work environment, for example strong social and communication skills, complementing those offered by men (see WEU 2001). However, in order to enable women to develop their skills and take part in the new economy by entering the workforce, I argue that the Government must examine the current support networks which help and enable women to maintain their domestic role as mother and carer at the same time as being a valued employee within the formal labour market.

Key characteristics of the new economy include the increase in working mothers, a higher proportion of lone parents at work and a rise in the availability of part-time employment. In this section I shall take each of these in turn and describe what changes have occurred, and subsequently set out what demands these groups might have of the new economy.

Increasing Working Mothers

As previously indicated there are now more working mothers than ever before. According to the Labour Force Survey, in spring 2000, 65 per cent of mothers with dependent children were in employment. There are many arguments proposed for the increase in women's participation within the labour market. For many mothers financial considerations play a major factor in decisions about their return to work, not just financial necessity to provide the basics for their family, but also, for some women, the need for financial independence. In 1998, 87 per cent of mothers in paid employment gave at least one financial reason for working, 35 per cent of full-time working mothers indicated that their main reason for working was due to a need for basic essentials compared with 48 per cent who are working to fill their need for extras (Bryson et al. 1999, 71). However, the benefits of returning to work are not always financial. For many women, earning an income brings with it self-esteem, and for mothers with low qualifications the need to socialise emerged as a more important reason for working (La Valle et al. 1999). For mothers with higher qualifications and job status there is often concern about the potential negative impact a long career break would have on their career status. In 1998, 76 per cent of women with high levels of qualifications (degree or higher) were active in the labour market when their youngest child was under five years compared with 27 per cent of women with no qualifications (Rake 2000, 28).

It can also be noted that women are returning to work sooner after giving birth than ever before. In 1996, 67 per cent of women who worked during pregnancy were back in work within nine months of having their child, compared to 24 per cent in 1979 and 45 per cent in 1988 (Callender et al. 1996, 152). According to Callender et al., 51 per cent of women who returned to work after having children returned into a lower occupational level and were paid less (1996, 163). Houston et al. (2000) carried out an extensive survey on 400 first-time mothers, before, during and after pregnancy. In pregnancy the women were asked to state whether they intended to stay at home with their babies, or return to full-time or part-time work at the end of their maternity leave. Before the babies were born, 20 per cent of the sample intended to stay at home, 26 per cent to work full-time and 54 per cent work part-time. Not all women carried out their intentions in relation to work. When their babies were one year old, only 82 per cent of those who intended to remain at home, 76 per cent of those who intended to work part-time and 62 per cent of those who intended to work full-time, were doing as they had hoped. The implications of this research are that, despite increases in the numbers of women who wish to combine motherhood with employment, the transition back to work is not easy. A greater level of support from both government and employers is needed to support women who try to combine work with motherhood to do so effectively. However, it is not always a financial or working hours issue. Many mothers believe that they are the best person to care for their children, and therefore choose to leave the labour market to bring up their child(ren). According to the Labour Force Survey, in spring 2000, 31 per cent of mothers with dependent children were economically inactive. For many non-working mothers, the perceived lack of family-friendly working arrangements played a major role in influencing their decision to stay at home. La Valle et al. reported that 66 per cent of mothers said that they would prefer to work or study if they had access to good quality, convenient, reliable and affordable childcare (1999, 4).

Lone Parent Families

With an increase in the number of divorces, a rise in the number of children born to non-married parents and the 'detraditionalisation' of the family, lone parent families are increasingly common within the new economy. In 1997, lone mothers headed 21 per cent of families, which is an increase from 11 per cent in 1984. This compares with 2.5 per cent and 2 per cent respectively for fathers (ONS 1997, 285). Lone parents often find it difficult to gain, and remain in, employment either due to unsuitable working hours or the salary not being enough to cover childcare costs. Between 1990 and 1997, the employment of lone mothers has risen slowly from 41 per cent to 44 per cent, yet according to Rahman et al. the UK still has one of the worst employment rates in Europe amongst lone parents (2000, 5). In its second term in office, the Labour Party set out a manifesto target that 70 per cent of all single parents should be in employment by 2004 (p26), and cuts in state benefits payable to lone mothers without employment were introduced to help reach this target.

Rise of Part-time Employment

One area of the labour market which demonstrates dramatic differences between men and women, is participation in part-time work (see figure 3.3): for many women, part-time work is seen as the ideal way to combine formal paid employment with motherhood (I shall discuss this further in chapter 6). Similarly to the EU, the increasing integration of women into the waged economy has coincided with the expansion of part-time employment within the UK. In 1999, 38 per cent of all working mothers with dependent children worked for 20 hours or less per week, compared with 2 per cent of fathers (EOC 2000a, 4). According to ONS (2000), 70 per cent of mothers choose to return to work part-time when they return from maternity leave. In 1996, first-time mothers who had worked full-time when pregnant were much more likely than those already with children to move into part-time employment after childbirth, 53 per cent and 19 per cent respectively (Callender et al. 1996). However, part-time jobs are often described as low skilled, low paid, precarious and insecure (Barron and Norris 1976, Hakim 1987). For part-time workers themselves this can mean a lack of job choice and status, as part-time employees are often seen as not fully committed to the job as they have other priorities at home, and employers can often bypass them for opportunities such as training and promotion.

Figure 3.3   Proportion of men and women of working age in part-time employment

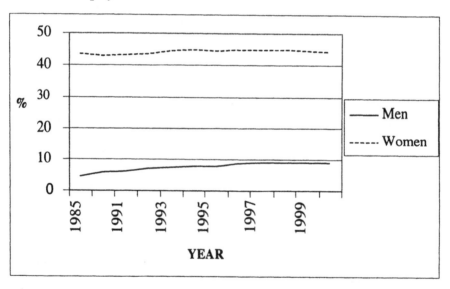

Source: Calculated from *Labour Market Trends* (June 2000 Table B.1)

With the increasing participation of women in the labour market and the recognition that women form an integral part of the new economy yet continue to maintain their domestic responsibilities, it is vital that the Government understand how and why women make the decisions they do. Whether women have the ability to choose to work or not work can be affected by Government policies. The next section of this chapter explores what the Government believes women want, and examines the policies the government is currently using to underpin their claim and to ensure women do have genuine choices as they move and take up their position within the new economy.

New Labour, New Female Labourers: Giving Women Choice?

Many women want to be able to make the choice of both going to work *and* looking after their families, rather than feeling they have to sacrifice one or the other. McDowell believes that the male breadwinner model of labour organisation that dominated the Fordist era recognised the significance of domestic labour, even though it depended on an inequitable gender division of labour organisation. However, recent moves away from the dominant male breadwinner role has placed the question of who is responsible for reproductive work and who is going to pay for it on the agenda of current social policy discussions (2001a, 457). The Government recognises that in addition to formal employment women are expected to take on additional domestic roles and consequently can face different and additional barriers to men in their choices around education, careers and families. Moreover a women's perception or interpretation of her role is as important a constraint on activities as any externally imposed restriction and this can prevent them from making the contribution that they want to make (WU 2000a). The Government believes that 'as many women who want to work should be able to do so' (Labour Party 1997, 25), yet it also values women's domestic role, and therefore understands that women need to be fully supported in whatever 'choices' they make albeit to (re)enter the labour market or to remain at home full-time in order to raise and care for their family.

> 'We value women in all the jobs they do whether at home or in the workplace. Strong families are the key to a successful society and women are the heart of the family. Supporting women in the role they play in those families is essential for any government. We believe that mums matter – society is indebted to everyone who has ever cared for and brought up a child.'
>
> (WEU 2000a, 1)

This quote from the Government asserts that women are entitled to a *genuine choice* about how much of their time they wish to dedicate solely to bringing up their family and that they shall be fully supported by the Government to enable them to fulfil their role satisfactorily. The Government aim to achieve this through the establishment of such initiatives as the Work Life Balance Campaign, which is encouraging employers to adopt family friendly policies and consequently increase access to employment for women. The National Childcare Strategy, which is to make improvements in childcare provision and other related policies such as the *Work and Parents: Competitiveness and Choice Green Paper* which has led to the development of a new Employment Bill. Each of these will be discussed in turn.

Work:Life Balance

The Government believe that most women do want to work,

> '8 out of 10 mothers work and 7 out of 10 say they would work even if they could afford not to.'
>
> (WU 2000a, 1)

On the other hand the Government also state that women perceive balancing work and domestic duties as increasingly difficult and so the government is keen to square the conundrum between home and work. In a consultation in 1999 with over 30,000 women 'they confirmed for us that that the biggest challenge facing them is striking a work:life balance' (Jay 2000a, 9). Even though women are keen to enter the workplace, a report commissioned by the Women's Unit in the Cabinet Office states that 83 per cent of women believe that commitment to family responsibilities hinders women's advancement in the workplace (Bryson et al. 1999, 6). This is supported by a time budget study, undertaken in 1995, which revealed that for every hour of paid work, a women's unpaid work drops by only half an hour. Interestingly, McDowell (2001a) points out that the higher women's own wages were, the more even was the gender division of domestic responsibilities within the household.

As Wilkinson argues, there is a serious tension for many between the demands placed on them to be good parents and spouses and to be high achievers in an increasingly competitive workplace (1998, 111). Contemporary political and policy debates acknowledge the need for encouraging a work:life balance, which I feel will only ever be fully achieved if the stereotypical gender roles of male 'breadwinner' and female 'homemaker' are destroyed and newly perceived roles of equal parental responsibility are created (the impact of these predetermined roles will be discussed in chapters 6 and 7). This role change will only begin to occur if the Government continues to strive to provide women with the choices which will enable them to succeed in their chosen career role. The Government pledges it is supporting women's choices and their desire to be successful at work as well as at home by working to promote the benefits of a work:life balance through the Work Life Balance Campaign. The Work Life Balance Campaign launched by Department for Employment and Education (DfEE) in March 2000 contained three main elements; firstly, the setting up of Employers for Work:Life Balance, an independent alliance of 22 leading employers committed in partnership to promote good practice in the community. Secondly, a new £1.5 million Challenge Fund (subsequently increased to £10.5 million) to help employers make work:life balance a reality by providing consultancy advice to forward-looking organisations that want to introduce appropriate working arrangements. Employers who make successful bids receive free consultancy advice about how to introduce flexible policies into their workplace. As the Government states,

'The Government is sitting down with employers, working hand in hand
to develop working practices that make a balance a reality for working
women. We have put our money where our mouth is and set up a fund
to help employers implement policies that help work life balance.'

(WEU 2000a, 2)

The National Assembly for Wales has established a separate fund of at least
£50,000 for projects in Wales. Thirdly, the Government published a discussion
document (see DfEE 2000b), which outlines the benefits for both employers and
employees of a family-friendly working environment.

The increased awareness of a need for work:life balance has become
particularly acute in recent years as our economy shifts towards a 24/7 service
economy. Consumers expect to be able to access services out of the normal 9-5
working hours and so business hours are expanding and consequently demands on
employees to be available outside of standardised hours are greater. 'Almost 19
per cent of employees work in workplaces operating 24 hours a day, seven days a
week. One in eight employees work on both Saturday and Sunday and non-manual
workers and those working in retail and transport are most likely to work a non-
standard week' (Hogarth et al. 2000, iv). As discussed earlier in this chapter, the
'out of hours' shifts suit some employees particularly well but others, however,
have difficulties working early or late shifts as they are unable to find childcare
during these times. According to Coote and Campbell, women need to work to
earn money, but they are busy. They have to look after their homes, their husbands
and children. They cannot work long hours or overtime, or awkward shifts which
interfere with cooking family meals or dispatching children to school. There are
periods in their lives when they need to work short hours; and there are times when
they can not work at all (1982, 63). It is having the option to negotiate hours
which is key to enabling women to (re)enter the work force. 'This government is
firm in its belief that flexibility is the key to a good balance between work and
home life' (Jay 2001b, 6). However, flexible working is not only about part-time
hours, it can include doing a full-time job during non-standard hours, or working
part of the time at home. Figure 3.4 sets out examples of various flexible working
arrangements which can be organised with the individual in order to help them to
carry out their employment satisfactorily at the same time as maintaining their
family role.

Figure 3.4   Options for flexible working

A -- Patterns that focus on how much time an employee works

- **Full-time work** - The employee chooses to work a typical 40 hour week.
- **Part-time work** - The employee works fewer than the standard number of hours a week, e.g. mornings, afternoons, school hours only, working alternate weeks.

   Types of part-time working may involve:
   - *Job-sharing* - Traditionally 2 employees share a full-time job.
   - *'V-time' working* - The employees works reduced hours for an agreed period at a reduced salary with a guarantee that he/she can return to full-time working when the period ends.
   - *Term-time working* - The employee does not work during school holidays.
   - *Working overtime* - The employee works more than the standard hours, in return for more pay.

B -- Patterns that focus on when employees can do their work

- **Flexitime -** Employees choose within certain limits when to start and end their working day. There are usually one or more periods of core time each day when the individual must be at work. Employees may be able to leave for a day or half day, if they have worked more hours than required; this is called 'flexi-leave'.

- **Compressed hours working** - This allows an employee to do a full-time job in, for example, four days a week rather than five. He or she may work 08:00 to 18:00 on Monday to Thursdays inclusive, and then have Friday off.

- **Annualised hours working -** Employees have to work a required number of hours each year. The hours worked each week vary throughout the year.

- **Shift working**

- **Shift-swapping** - Two employees work each other's shifts, so that one or both can attend to other concerns when they would normally be at work.

- **Working outside normal hours** - Some organisations – such as hospitals – have to keep going at nights, on Sundays or on public holidays. So they expect at least some employees to work at such times. Employees accept this as an integral part of their job. Some may even prefer to be at work when their partner is at home or at times that attract premium rates of pay.

- **Self rostering** - Staff can choose the patterns they want to work, within agreed parameters, while meeting the needs of the business.

C – Patterns that focus on where employees work

- **Working at the employer's premises** - Employees often have to work at their organisation's office or factory, for example, if they provide a face-to-face service for customers at a branch office, or they work on a production line. Even where the work does not have to be done at the employer's premises, many employees prefer to work from there for a variety of reasons.

- **Working from home** - Employees do some or all of their work from home. This is most successful when there are good communication links between home and the office, and when the employee does not need close supervision. Types of work that can be especially suited to home-working include sales and marketing, editing, accounting and providing a telephone answering service.

D – Patterns that give a complete break from work

- **For a short period** - Examples are paternity, maternity and adoption leave, where parents take time off around the time when a child is born or adopted, and parental leave, which they can take in the years following the birth or adoption.

- **A longer absence** - Examples are an unpaid career break or a paid sabbatical.

E – Packages that offer choice and security to employees

- **Company benefits** - such as childcare or eldercare vouchers

- **Phased or flexible retirement**

Source: Adapted from DfEE 2000b, 15-16

Flexibility appears to be growing amongst employers as they recognise the benefits to them, which include reduced casual sickness absence, improved retention, improved productivity and improved morale and commitment (DfEE 2000a). However, employees with children are unable to work flexibly unless they have the support either from the welfare state or friends and family. As discussed in the next section, the Government is working hard to make improvements in the type and availability of childcare in order to support women in going back to work

Childcare Strategies and Their Contradictions

The percentage of three and four year olds in maintained nursery and primary schools in England has increased from 44 per cent in 1987 to 56 per cent in 1999 (EOC 2000a, 4). Yet, for every seven children aged under eight there was only one place in a day nursery, with a registered childminder or at an out-of-school club (DfEE 2000). Despite this, Bryson et al. (1999) argue that only 33 per cent of mothers who do not work cite the lack of suitable childcare as the reason. However a recent analysis of the evaluation of the New Deal for Lone Parents showed that more than half the lone parents taking part in the study cited problems with finding childcare as the greatest barrier to entering employment (Britton 2001). On their research on the demand for childcare, La Valle et al. (1999) reported that nearly three-quarters of parents in their sample said that their childcare arrangements were not ideal, with the most commonly cited reason for their dissatisfaction being lack of local provision and inability to afford more adequate childcare. In the ideal world where parents were not constrained by availability or cost, an overwhelming majority said they would prefer an informal provider (for example relatives and friends). According to Riddell (2001), there is only currently one subsidised childcare place for every 14 children in poverty and although the Government has promised to introduce 100 centres of excellence and 900 neighbourhood nurseries in deprived neighbourhoods by 2004, Riddell argues that it has been estimated that 10,000 children's centres, offering pre- and after-school care are actually needed, at the cost of around £10 billion (p28). Very few employers offer any assistance with childcare to their employees. Only two per cent of employers provide a workplace nursery and only one per cent subsidise nursery places (Hogarth et al. 2000).

Women who return to work after having children need to have someone reliable to look after their child(ren) and know that their job is going to be flexible enough to allow them to carry out family duties when necessary. The Government claims to understand this (see WEU 2000a) and feels that it has already implemented policies to this end, such as the National Childcare Strategy, whose aim is to ensure good quality, affordable childcare for children aged up to 14 in every neighbourhood, including both formal childcare and support for informal arrangements. With only 830,000 registered childcare places for the 5.1 million children under the age of eight in England, the National Childcare Strategy began in May 1998 with the publication of *Meeting the Childcare Challenge* Green Paper. The National Childcare Strategy promises that between 1998 and 2003, £470 million will be made available in England to meet its target of creating one million accessible and affordable childcare places by 2003. The Government believes that by investing the money and improving the provision of childcare, women will be able to (re)enter the labour market and use their skills in the new economy more easily. Whether or not the solution to the barrier of women entering the labour market will be provided by improvements in childcare provision will be discussed later in this book. The Government, however, continues to claim that,

> 'In the new economy, provision for childcare will be seen not as a cost, but as an investment. Why? Because making sure that women can benefit from high quality childcare liberates them to use their skills to everyone's benefit.'
>
> (Jay 2001b, 9)

The key points of the National Childcare Strategy are quality of childcare, affordability of childcare, accessibility of childcare and partnership. I shall explore these in turn.

The National Childcare Strategy acknowledges that the quality of childcare can be variable, and depends upon well-trained, competent staff to provide a safe, stimulating and enjoyable environment. Funding has therefore been allocated specifically for Early Years Training and Development, and all early-years settings in the maintained, private and voluntary sector who receive Government nursery funding, must be inspected by The Office for Standards in Education (Ofsted) to ensure they are delivering good quality education for children. Since September 1998 the Government has provided free part-time nursery places for every eligible four year old whose parents want it. This entitlement is for five 2½ hour sessions per week. Since April 1999, Local Education Authorities have had to ensure that there are sufficient places for all eligible four year olds in their areas.

Estimates from the Daycare Trust (2002) state that the current cost of placing a child under five years old with a full-time childminder can range from £50 to £120 per week, depending on demand and location. Overall this is the most expensive in Europe. However, when a child starts school, costs do not disappear and parents are faced with paying between £15 and £20 a week for an after-school club and between £50 and £80 per week for a holiday play scheme. La Valle et al. reported that 36 per cent of respondents said they found it difficult to pay for childcare. Respondents reported that with a 25 per cent increase in costs, a fifth of respondents said they would have to reduce or stop using childcare altogether. A potential decrease of 25 per cent in costs would encourage over a third of parents to get more childcare, with 12 per cent saying they would use the additional 'free' time to do some learning or training and 13 per cent to work more hours or look for another job (La Valle et al. 1999, 5). The Government are helping parents with the cost of their childcare through the introduction of Working Family Tax Credit and the Childcare Tax Credit. The Government explains that 'Our vision of the tax and benefits system for families with children is to provide help for all families; to give most help at the time families need it most; and to give more help to those families most in need' (Labour Party 2001, 25). Working Family Tax Credit is a tax credit payable to working families with low to middle incomes depending on circumstances, and is paid through the pay packet. It guarantees an income of £207 per week for all families with a parent in full-time employment. The Childcare Tax Credit is worth 70 per cent of eligible childcare costs up to a maximum of £70 a week for families with one child and £105 for families with two or more children. Eligible childcare is that which is registered with the local authority, such as registered childminders, play schemes, day nurseries or out-of-school clubs.

There is a lack of information about what childcare options parents have locally, and so Early Years Development and Childcare Partnerships have been asked to set out a local information strategy showing how services will be developed so that all parents can find the information they need. The National Childcare Strategy acknowledges that:

> 'At present availability of childcare is a matter of geography, not need. We aim to put this right. We want to improve information for parents in what childcare is available. And we want to ensure a diversity of services to meet parents preferences – from childminders to day nurseries and including private, voluntary and local authority run services...We want this to be consistent across the country.'
>
> (1998, 4)

The recognition of the geographical variation in the provision of childcare across the UK is key to understanding the ways in which the National Childcare Strategy needs to be implemented, for it will vary with location. In order to find out all childcare options, the Government says that women will be able to ring a national helpline to find out what is available in the area they live (WEU 2000a, 3). However, La Valle et al. reported that a large proportion of parents (44 per cent) had not used any sources of information to find out about local childcare provision with 39 per cent reporting word of mouth as the most important source, (1999, 3).

The Government believe that working in partnership is the key to the successful delivery of the National Childcare Strategy. The partnerships consist of the Early Years Development and Childcare Partnerships, set up in every local education authority in England and Wales. Representatives are from a range of organisations covering private, voluntary, maintained sectors, parents, educational organisations, health services and information services. The partnerships are designed to be inclusive in formulating local childcare strategies. Overall, the Government promise that through the implementation of the National Childcare Strategy 'By 2004 we will have created childcare places for 1.6 million children. That is on top of the universal free early education provision we have already made for all four year olds whose parent wants a place, and will extend that to three year olds by 2004' (Jay 2001a, 8).

The Government reported in May 2000 that 120,000 new childcare places were created in 1999, exceeding targets (Britton, 2001, 1).

The National Childcare Strategy for Wales follows the same remit as that delivered for England, with a total of £7.3 million available to take forward and deliver the Strategy. A Task Force has worked alongside the Welsh Assembly Government's Early Years Advisory Panel, which is implementing the Assembly's commitment to make a free early education place available for all three year olds, for half the day in term time. The Welsh Assembly Government (WAG) believe that 'our economy will prosper if more skilled and capable people are able to take up opportunities because they have access to good quality, affordable and accessible childcare' (1998, 8). The Welsh Taskforce has reported to the WAG on the current status and issues for implementation of the Childcare Strategy in Wales. The taskforce reported that the cultural emphasis of stereotypical male and female roles and responsibilities has restricted the growth of full-time day care and costs have also been prohibitive, other than for professional couples. This has led to a heavy reliance on the voluntary sector and, most significantly in Wales, to a reliance on family members, relatives and or neighbours through informal arrangements. The National Childcare Strategy Taskforce believe that improving childcare provision within Wales will not only support parents in (re)entering the labour market, but it will also influence the educational and economic regeneration of the area.

> 'There is a growing consensus that, until local communities throughout Wales develop appropriate services for children, including those that can support parents in work or training, then little impact will be made to the status quo...A strategy that places the child's right to access enriching experiences as the central tenet will have a knock-on effect by also supporting the educational, economic regeneration, community development and employment outcomes, whilst removing the current disparity in access based on *parental* (rather than children's) need and "postcode lotteries".'
>
> (National Childcare Strategy Task Force 2001, 6)

The National Childcare Strategy for both England and Wales recognises a geographical disparity in the availability and quality of childcare provision. The National Childcare Strategy Taskforce for Wales reports 'current funding for childcare settings and voluntary sector services is grossly inadequate, multi-sourced and not uniform across local partnership areas and within local communities in Wales' (2001, 13). In a report for DfEE, La Valle et al. (1999) reported the highest levels of use were in the South West and the lowest in London. La Valle et al. linked these results to the proportion of people with very low incomes, and households where the mother was not in employment, which were greatest in London and least in the South West. Issues of regional disparity in childcare usage will be explored further in the next chapter. Further analysis of La Valle et al.'s results highlighted that the strongest predictors of childcare use were child's age, household structure and the number of children in the household.

In addition to UK Government initiatives such as the Work:life Balance Campaign and the National Childcare Strategy, the Government are also encouraging the establishment of new legislation to provide support for working parents. Although new legislation may not be directly aimed at women, in this section I shall draw out the impact which the new changes will have for working mothers. For example the Working Time Directive limits working hours for many people to 48 hours, therefore helps to ensure that employers are limited in the demands they can place on their employees, and ensures parents have at least a minimum right to time off to be with their family. The Part-Time Work Directive helps parents who wish to reduce their working hours to do so, without losing their employment rights. On 1st July 2000, the Government introduced new rights for part-time workers which ensure that Britain's six million part-timers, most of whom are working mothers, are not treated less favourably than comparable full-timers in their terms and conditions, unless it is objectively justified. This means part-timers are entitled to:

- the same hourly rate of pay,
- the same access to company pension schemes,
- the same entitlements to annual leave and maternity/parental leave on a pro rata basis,
- the same entitlement to contractual sick pay, and
- no less favourable treatment in access to training.

As previously discussed, many women who returned to work after maternity and reduced their hours to part-time experienced occupational downgrading, if not literally, then submissively through being passed over for promotion or training. The Part-Time Working Directive measures, reinforce the Government's policy of putting in place minimum standards whilst promoting a flexible and competitive workforce, which are useful foundations for the Government's wider family-friendly employment policies.

More recently, and more focused on parental needs, was the publication of the *Work and Parents: Competitiveness and Choice* Green Paper (Dec 2000). The proposals set out in the Green Paper were developed after other research revealed to the Government the following: mothers were dropping out of the labour market because they had to come back from maternity leave too soon; fathers are wanting to be more involved with their new babies; both mothers and fathers are wanting to work more flexible hours, and there is a need for more flexible and extensive childcare (DTI 2001). In two-thirds of families both parents work outside the home, and the Green Paper set out 'to review the steps needed to make sure that both parents have choices to help them balance the needs of their work and their children, so that they may contribute fully to the competitiveness and productivity of the modern economy' (p5). The Green Paper aims to provide families with options for genuine choices: to look after their children full-time, or to combine work, education or training with parenting in a balanced way. When choosing whether to return to work, women want to know that they will be able to deal with the crises that inevitably arise in family life. One of the main issues which the government has discovered is that in some cases, for example in the case of family emergencies, although all parents already have a right to unpaid time off, both employers and individuals are often unaware of it (p25). The open consultation for the *Work and Parents: Competitiveness and Choice* Green Paper closed in March 2001, and after observing responses, it formed part of the Employment Bill introduced to Parliament 8th November 2001.

Employment Bill

Fathers have had the right to unpaid parental leave since 1999; however, the Bill provides for the introduction of a new right to two weeks' paternity leave, in addition to the current entitlement of 13 weeks' parental leave following the birth of a child. This provision will come into effect from 2003. Statutory Paternity Pay (SPP) will be £100 a week or 90 per cent of average weekly earnings, whichever is the lower, and will be administered by employers in the same way as Statutory Maternity Pay (SMP). The Government will finance paternity leave, with employers able to recover the majority or the entire amount that they pay out.

Additionally, the Government has discovered that women entitled to longer periods of maternity leave are more likely to return to work. Seventy-two per cent return after 28 weeks leave, compared to 50 per cent after 18 weeks. Some women feel distracted and less productive when maternity pay ends, and there is concern among early returners that they have to give up breastfeeding too early for their child's well-being (DTI 2000, 16). Women returning after a longer period often feel better prepared. In the Green Paper, the Government consulted on whether a negotiation of extended maternity leave is an area which should be left to best practice or whether they should legislate to give parents, at the end of the maternity leave period, the right to opt to work reduced hours provided it did not cause harm to their employers. The Work and Parents Taskforce (WPT), which was established to look at how best to implement the legislative right for parents of young children to request flexible working hours, reported that the results were polarised; parents and their representatives wanted legislation while employers opposed it. In general the WPT are supportive of the governments proposals, however, and are keen to indicate that a varied and comprehensive package of support is necessary to accompany the duty to consider requests to work flexible hours: they point out that one guidance booklet, as currently provided, will not be enough. The WPT also suggest that the Government sets up challenge funds to give employers access to advice at a regional level, possibly through Business Links or the Regional Development Agencies. In response the Government has accepted, or accepted in principle, each of the WPT recommendations. The Bill provides for an increase in the maternity payment period from 18 to 26 weeks, allowing a woman to stay at home for a year in total. The Bill also provides for an increase in standard rate of payment for Statutory Maternity

Allowance from £62 to £100 a week, or 90 per cent of average weekly earnings, whichever is the lower.

The Government also intends to give working parents with children under six years who have been with their companies for a minimum of six months, the right to make a written request for flexible working, although companies can reject the request, they must set out a considered business case for doing so.

Extending parental leave was not offered as an option in the Green Paper, but was raised in discussions after its publication and so the Government felt it should be addressed. It subsequently announced its plans to make changes on 25th April 2001. The new right to parental leave applies to employees (both mothers and fathers) who have completed at least one year's continuous service with their current employer. The leave is an allowance of 13 weeks' parental leave for each child which can be taken anytime (only in blocks of a week, maximum of four weeks a year) up till the child's fifth birthday. At the end of parental leave, the employee is guaranteed the right to return to the same job as before, or if that is not practical, a similar job which has the same or better status. Parental leave will remain unpaid. According to Wilkinson, paid parental leave is one of the few policy tools with a proven track record in helping to reverse declining birth rates. Scandinavian countries have seen a dramatic return to family formation since the 1980s as a result of generous paid parental leave (2001, 123). Although through some of its initiatives the Government may appear to want to get women into the workplace, the current legislation consultations do indicate that it is aware of some of the barriers that women face in doing so. However, the Government is making assumptions about what women want and need based on national economic figures, which may not represent what is happening at the local level. Suggestions for policy makers to consider will be made in chapter 8.

Contradictions within Labour's 'New Economy'

As we experience a 'detraditionalisation' of the workforce and move towards a 24/7 economy, we are in need of new support networks in order to enable women to continue their domestic role at the same time as maintaining their position within the labour market. According to Giddens, the family figures in some of the most contentious debates in modern politics. However, 'sustaining family life cannot be achieved through a reactionary stance, an attempt to reinstate the "traditional family" ...it presumes a modernising agenda of democratisation' (1998, 18). Politics and policies need to appreciate the shift in the twenty-first century towards a post-divorce society (where divorce is commonplace), the post-marriage society (where the role and function of marriage is increasingly called into question) and the growth of childfree families (Wilkinson 2001, 118). The Government argues that it is currently helping to promote a better work:life balance in the following ways:

- Working with the public and private sectors to promote work:life working policies and practices.
- The National Childcare Strategy will increase access to good quality affordable childcare – providing one million new childcare places over the next four years.
- Introducing the Working Families Tax Credit and the Childcare Tax Credit.
- The Parental Leave Directive, allowing working parents of children under five to take up to 13 weeks' leave, and have the right to time off to care for sick children.

(WU 2000a, 3)

There has been a general acceptance of the proposals which have been put forward by the Government, although individual responses have highlighted different concerns. The Industrial Society for instance, one of the UK's leading think-tanks and advisors on the world of work, commend what the Government is aiming to achieve, particularly in its recognition of the role of working fathers. However they argue (2001) that the changes which are proposed are the minimum which is necessary and they will continue to lobby for further paid paternity leave, paid parental leave and the right to have a request for flexible working hours considered fully and fairly. However, not all responses are as positive as the Industrial Society. Sixty-five per cent of respondents to a survey of 500 members of the Institute of Directors (IoD) expressed their opposition to the introduction of the paid paternity leave. The IoD predict that some businesses, particularly small businesses, could experience organisational problems occasioned by the absence of staff on paternity leave (IoD 2001, 4). Employers would either have to take on temporary employees to cover for staff absent on paternity leave, or increase the burden on other full-time workers. With regard to maternity rights, The IoD feel that the proposed measures are likely to increase business costs because employers may find that they have to arrange cover on the absence of staff taking maternity leave. In summary, 'the so-called family friendly policies that feature in the Employment Bill are not popular with IoD members' (IoD 2001, 4). In contrast, Wilkinson argues that the present 'family-friendly' employment agenda is not radical enough: differentiating between workers with families and those without can lead to an unhealthy tension between the two, creating more obstacles to the much-needed shift in working styles and practices in the mainstream culture of work (2001, 123). The Industrial Society suggest three key tests which any proposals on work and parents should pass; first, whether they actually increase the overall amount of time available to parents; second, whether the proposals reduce the differentiation between mothers and fathers and finally whether the proposals increase flexible working (2001, 2).

Reeves (2000) argues that unless policies emphases 'parental' and especially 'paternity' rights to leave, in order to break down the roles and responsibilities attached to fatherhood and motherhood, women will continue to be discriminated against as employees. Flametree, a think-tank which specialises on workforce issues and consults with organisations on how to respond effectively to the work:life challenge, also express their support for the proposals in the significance they provide for fathers who will be able to request flexible working and working parenting will no longer be viewed as the principle domain of the mother (Flametree 2001). However, I feel that although it is a start in the right direction, it will take more than the implementation of a new Employment Bill to change the engrained stereotypes of man as breadwinner and woman as homemaker. As Wilkinson argues 'the emphasis has to be in fostering change in values and attitudes' (1998, 113). Wilkinson (1998) and Giddens (1998) believe that only a minority of children now grow up in a traditional family, that of the male breadwinner, female homemaker and two children. Consequently, Wilkinson argues that this is, therefore, far too limiting a model for family policy and believes that the emphasis should be on ensuring that working families benefit from high quality subsidised childcare, from varieties of paid parental leave, and from family-friendly tax and the benefit system. Wilkinson believes that 'Government will look to employers to play more of a role on facilitating changes in the culture of workplaces. Public-private partnerships will become commonplace as the government seeks to place a share of financial responsibility for family life on tax payers and employers, in recognition of the fact that individual parents have paid too high a price in recent decades in foregoing an advancement at work to take on the role of childrearing' (Wilkinson 1998, 116).

It is worth noting that three out of the four ways in which the government aims to help promote a better work:life balance involve childcare. The Government promotes the importance of parents having genuine choices, yet the policies and administrative guidelines are actually set to coerce women into the workforce. Rather than encouraging mothers to take responsibility for childcare, it is preferable to encourage them to stay in employment (Esping-Andersen 2000, 758). The problems being experienced by families today are rooted in both economic stress and in family disintegration. Wilkinson argues that any progressive family policy must address both these issues or it will fail (1998), something which the Government also acknowledges. According to the Minister for Women the Government needs

> '...to move on to look at the relationship between this infrastructure of support to recognise the changing nature of family life, and its connection with work and family policy. There is I think a dilemma here. On the one hand we talk about the importance to the economy of working and on the other hand we talk about choice.'
>
> (Jowell 2000, 11)

McDowell (2001a) argues that in most cases the British Government has taken a minimalist approach to new policies, restricting the take up of new rights such as parental leave to as small numbers as possible. Flametree believe that the shortcoming of the proposals is that only parents of children under six will be eligible, although childcare and the child's needs for the over sixes can be equally demanding (2001, 1). McDowell (2001a) argues that at present British rights to parental leave are the least generous in the EU and states that Britain was also one of the last members of the European Union to introduce equality between full-time and part-time workers in terms of rights to leave and holiday entitlement. Parents at Work, a charity which provides advice and campaigns on behalf of working parents, argue that the present low level of maternity pay means that many mothers are forced to return to work before they feel able to cope, and therefore approve of the Government's proposals to increase and extend maternity rights. Research shows that the majority of parents cannot afford to take unpaid parental leave (Parents at Work 2001).

As previously mentioned childcare is often seen as *the* barrier stopping women (re)entering the workforce, and is often at the heart of the struggle to find a work:life balance. However, I feel there are many more influencing factors, and women's participation in the labour market is more than simply a childcare issue. I feel it is engrained further in the continued existence of stereotypical images of 'breadwinner' and 'homemaker', childcare issues and the local economy. The national picture drawn by the Government is interesting; however, it does not reveal much about how individual women are reacting to the pressures on them from the new economy. Academic work, as previously discussed, points to the importance of geographical variation, stressing the need to see labour markets as locally constituted entities. For example, through the development of fourth generation segmentation theory, it becomes evident that there is a need to explore how individual women are balancing the complex demands of work and home in the locality. This will enable an investigation of whether it is in fact possible to have genuine choices within a competitive economy, and whether those choices might vary from place to place. In the rest of this book I will explore whether geographical variation is a major issue affecting women's 'choice' about (re)entering the labour market.

Conclusion

Through an examination of women's position in the UK labour market, this chapter has demonstrated that women are an important part of the new knowledge-based economy. Moreover, this chapter has also demonstrated that although the Government acknowledges women's vital contribution to the labour force, it is not recognised sufficiently in national labour market policy. Due to the way in which Government policy is formulated (using nationally aggregated data) there is a distinct lack of acknowledgement for local variations, and a blanket policy is consequently applied to the whole of the UK. This has led to a concentration on particular individual barriers such as childcare. As discussed earlier in this chapter, partnerships have been established between childcare providers, but no other barriers to women have been addressed in the same manner. I shall now go on to examine the degree of spatial variation in women's labour force participation, which exists within the UK which is not represented by a national analysis.

Chapter 4

The Uneven Development of Women's Participation in the Labour Market

Introduction

> 'While the nation state is both an important actor and scale for action, this neglects other spatial scales which may be of central importance to how social processes operate.'
>
> (Duncan 2000, 16)

Through an examination of the academic literature on the position of women in the labour market, including the development of fourth generation segmentation theory (see Peck 1996a, Kreckel 1980), the impact of household interdependencies in women's decision-making process (see Hanson and Pratt 1991, 1995), and an analysis of UK national data, I have so far argued that a national scale of analysis could conceal important regional and local variations of women's participation in the labour market. In this chapter I shall further argue that national analysis alone cannot explain or explore the processes and practices of what constitutes the local labour market, and that it is the composition and construction of varied local labour markets, which in turn helps to explain women's participation in the labour market. This chapter questions whether nationally aggregated data conceal important variations at the regional and local level, and posits that any clear understanding of women's participation either within, or outside, the formal labour market, must come from an investigation of women's decision-making process and an understanding of the complex networks within which women operate.

This argument is developed in three parts; firstly, I shall describe the current level of variation in women's participation within the UK labour market, and point out that this is not simply a reflection of the status of the local economy of that area. Secondly, I shall argue that only by looking at the many processes which construct the local labour market can we begin to gain a full understanding of women's labour market participation. This takes labour market analysis beyond concerns with national data. Thirdly, I shall examine some attempts by geographers to integrate spatial variation into labour market theory, arguing that the majority of attempts have insufficient sensitivity to the importance of social factors in their typologies. Moreover, categorised typologies do not provide an explanation for the complex network of processes which operate at the local level and fail to provide explanations for women's labour market participation. Finally I shall consider whether an examination of the local labour market provides a complete explanation for variation in women's participation. Peck argues that labour market participation is profoundly shaped by the networks of spatial and temporal routines which women must construct between waged work on the one hand, and various domestic and caring tasks on the other (1996a, 334). Additionally, I shall question whether, in fact, an exploration at the regional or local level is enough to provide an understanding for the variation in women's labour market participations, or whether we need to explore the daily lives and complex household interdependencies of individual women which operate within and around their household unit.

The Spatial Variation of Female Labour Market Participation in the UK

There are several sources that provide national aggregate data on women and employment, many of which were mentioned in the previous chapter. These include government sources such as The Women and Equality Unit within the Cabinet Office, the Department for Education and Skills (DfES), Office for National Statistics (ONS), government-sponsored initiatives such as the National Childcare Strategy and The Work-Life Balance Campaign, and non-government organisations such as The Equal Opportunities Commission and Working Parents. As was pointed out in the previous chapter, due to their different methods of data collection, every resource provides differing results, which consequently causes contradictions when aiming to obtain an accurate national picture of women's labour market activity. Perhaps the most reliable and comprehensive source of national data in the UK is the Labour Force Survey (LFS), administered by the government's Office of National Statistics. The LFS is a continuous, household-based survey which provides a wide range of data on labour market statistics and related topics, such as occupation, training, hours of work, qualifications, income and disability. The data from the survey is used extensively both within and outside government. The LFS is carried out under a European Union Directive and uses internationally agreed concepts and definitions, which is the source for the internationally comparable (International Labour Organisation) measure known as 'ILO unemployment'. The UK is required by European Union Regulation to carry out a labour force survey annually. Results from the spring quarter of the LFS are supplied to Eurostat each year to meet this requirement.

Each quarter, LFS samples 60,000 households, receiving approximately 138,000 respondents based on a systematic random sample design, which makes it representative of the whole population of Great Britain. The survey seeks information on respondents' personal circumstances and their labour market status during a specific reference period, normally a period of one week or four weeks (depending on the topic), immediately prior to the interview. The smallest unit of data collected by LFS is local authority district. No other regularly, nationally collated data on labour market participation can be broken down to local level. Although UK census data may provide greater coverage and in-depth analysis, there are questions to be asked about its reliability, as it is a self-declared dataset. Additionally, at the time of investigation for this book (spring 2000) the latest census data available was 1991, which would not be representative of current labour market activity.

The survey sample of addresses is taken from the Postcode Address File. In addition, a small sample of addresses of NHS and Health Trust accommodation is included in the survey and anyone aged 16 or over and at boarding school or living in a hall of residence is included in their parent's household. Throughout the UK 0.2 per cent of households are sampled, which is made up of a 0.2 per cent sample from each local authority area. The survey has a stratified random sample and within any continuous thirteen-week period every postcode sector is sampled. This feature allows representative results to be produced for any thirteen-week period. Given the size of the sample, the considerable variety of population characteristics within local authority areas and the fact that such areas do not correspond with local labour markets, there are likely to be significant random errors in the degree to which the data represent each local area. Nevertheless the LFS data enables me to compare local authority areas and to identify broad patterns in the participation rates of women with children. An examination of the variation of women's participation in the labour market can be seen in figure 4.1 which demonstrates the range of the variation from the UK national mean of 69 per cent.

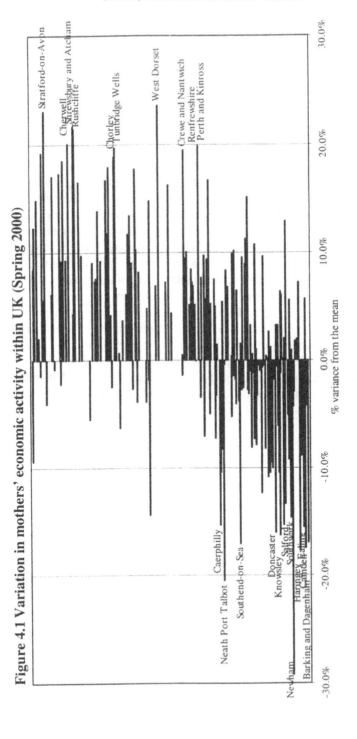

Figure 4.1 Variation in mothers' economic activity within UK (Spring 2000)

Figure 4.1 clearly shows a wide geographical variation of mothers' participation within the formal labour market in the UK. There are only a few districts close to the national mean including Rochdale, Stockport, Wolverhampton, Derby, Bristol, Southampton and Swansea. However female economic activity ranges from 40 per cent in Newham, an east London Borough, to 93 per cent in West Dorset, a district on the south coast of the UK. One of the key issues this book addresses is why we have such a wide variation of women's participation in the UK formal labour market. Forsberg (1998), Duncan and Smith (2002) and others (see previous chapter) have argued that debates about the position of women within the labour market, tend to be restricted to the national level, or make comparisons between different welfare state regimes (see Esping-Anderson 1990 and Sainsbury 1994). Duncan and Smith believe that although this body of work has provided an important contribution to understanding female labour market participation, it can be criticised for focusing on the national level, and neglecting important regional and local dimensions to gender inequalities and family behaviour (2002, 22). Is it that the women in high economic activity areas such as West Dorset and Stratford-upon-Avon, are choosing to work and women in low economically active areas such as Newham and Neath Port Talbot are choosing to stay at home? Or do women in Newham and Neath Port Talbot not have the opportunity to work, and those in West Dorset and Stratford-upon-Avon do? Is it a choice of many options, or a 'choice' from a lack of options? This book questions whether these statistics are a reflection of the 'choices' women in such areas have made, or asks whether there are other factors which have forced them into a particular lifestyle. In order to answer such questions we can agree with Fosberg, who argues that 'if we want to analyse welfare from the point of view of people's everyday lives, then we must also go to that level where welfare is produced and consumed – the local level' (1998, 193). In this chapter I suggest that the answer to these questions will only come from an understanding of the complex system of interdependencies, which underlie women's daily lives at the local level.

Prior research has argued that high female economic activity is associated with areas of similar economic trajectories, especially in areas of declining, traditionally male industries, such as mining and shipbuilding: for examples see WGSG 1984, Green and Owen 1998, Cooke 1983a. These will be discussed in detail later. In these declining economic areas, women have arguably either been forced into employment to compensate for heavy male job losses: alternatively service industries, which are argued to be more suited for female employment, have moved into the area proactively recruiting a female workforce (see Massey 1984). However, as can be seen from figure 4.2, the types of areas with the highest and lowest female participation rates are diverse and it appears that there are a number of different processes, operating at a local level, which cannot be categorised and related purely to the economic history of an area.

**Figure 4.2   Economic activity rates of mothers with dependent children by
local authority district level: top 10 and bottom 10**

Areas with highest female economic activity rates	%	Areas with lowest female economic activity rates	%
West Dorset	93	Newham	40
Stratford-upon-Avon	92	Neath Port Talbot	49
Cherwell	91	Haringey	51
Rushcliffe	90	Camden	51
Shrewsbury and Atcham	89	Southend-on-Sea	52
Perth and Kinross	89	Barking and Dagenham	52
Renfrewshire	89	Knowsley	53
Tunbridge Wells	89	Doncaster	53
Crewe and Nantwich	88	Ealing	54
Crawley	88	Caerphilly	54
Chorley	88	Salford	54
		Southwark	54

Source: *Labour Force Survey*, Spring 2000

It is, therefore, not possible to say that those communities with high female economic activity are old industrial/mining communities with new developing service economies, for at either end of the scale we have a wide range of local economies. At the high end of female economic activity lies West Dorset which is a tourist and seaside area, Stratford-upon-Avon which is a historical tourist area and Cherwell and Rushcliffe which are ex-mining and service economies. At the lower end of mothers' economic activity we have Newham, which is an outer east London Borough with some commuting, Neath Port Talbot, which is an ex-mining and industrial area and Southend-on-Sea, which is a seaside town with tourism. These huge local variations in women's labour market participation, which are possibly not determined by the presence or absence of particular local economies, are disguised by reference to the 'national' female economic activity rate of 69 per cent.

Other research, which supports the lack of association between local economic histories and the increase of female participation in the labour market, has come from Duncan and Smith (2002) who use the notion of a Motherhood Employment Effect (MEE) to show the importance of regional variations, within any given national average. The MEE is a standardised measure of the difference between the full-time employment of partnered mothers (i.e. with dependent children) and the full-time employment rate of non-mothers (i.e. without dependent children) in the prime 'motherhood' age range of 20-45 years (Duncan and Smith 2002, 23). In an analysis of the UK, the MEE range varied from 50 per cent in Lancaster and inner London, where almost half of all mothers withdrew from the worker role, to almost 90 per cent for outer suburban towns in the southeast of England, where almost all mothers withdrew from the worker role. Duncan and Smith therefore point out that the broad regional patterns of the MEE do not correspond with patterns of economic growth and decline. For example, both Lancaster (an area of relative economic stagnation) and west London (with better job prospects for women) show low withdrawal rates, indicating low adherence to the male breadwinner/female homemaker stereotypes. Similarly, Duncan and Smith highlight that those areas with the highest employment growth in Britain (the outer southeast and East Anglia) and those with the highest job losses (the former coalmining and steel-making areas in South Yorkshire and Nottinghamshire), both show high withdrawal rates and hence higher adherence to the male breadwinner/female homemaker norms. Nor does the MEE rate correspond with the urban-rural differences. Some large towns show high withdrawal rates (Bristol, Hull, east London and Sheffield), as do some of the more remote rural areas in Cornwall or Galloway in southwest Scotland.

The lack of correspondence between women's labour market participation and broader economic indicators is also supported by Jarvis (1997) who looked at the distribution of 'traditional' households (men in full-time work, women as full-time housewives) across the UK. Perhaps contrary to expectation, Jarvis found that 'traditional' households are most common in high growth areas in the southeast of England, while low growth Lancashire and Greater Manchester, have the lowest proportion of traditional households and the highest rates of dual-earner households. Overall, the evidence provided in figure 4.2, along with the work of Duncan and Smith and Jarvis, shows a lack of connection between high economic growth areas and women's participation in the labour market. Neither the simple north-south dichotomy nor the more detailed labour market regionalisations, which are both based upon economic indicators of growth and prosperity and on social indicators of class and well-being, show much correspondence with the geography of gendered divisions as shown by the mapping of the MEE and the distribution of 'traditional' households.

Such literature, stretching back some twenty-five years, leads us to ask what does the national mean of 69 per cent actually demonstrate in relation to this local picture? And perhaps more importantly, what does a national mean of 69 per cent conceal when further exploring the variation within local labour markets? Through an understanding of the huge variation of women's participation in the labour market at the local level, Duncan and Smith's work on the MEE and Jarvis's research on the distribution of 'traditional' households, which confirm the lack of correlation between the status of local economies and mothers' labour market activity, we must accept that the existence of local institutional networks play a significant part in explaining the variation of women's participation in the labour market. In the next section of this chapter I shall explore the ways in which typologies have been applied to local labour markets in an attempt to group together the processes at work which influence the constitution and production of different types of labour markets in different places.

Exploring Variations: Local Labour Market Typologies

> 'It is not possible really to understand the geographical pattern of
> unemployment decline without going behind the pattern to an
> understanding of the processes creating job loss.'
>
> (Massey and Meegan 1985, 119)

If as suggested by Massey and Meegan, we accept the warning about the
importance of moving from pattern to process, we must also be careful about over-
simplifying the processes. Bennett and McCoshan remind us that

> 'Generalising across different situations is clearly subject to a high level of
> local variability.'
>
> (Bennett and McCoshan 1993, 223)

There is no denying that the national picture is important in providing a broad
reference point, especially for government policy. However, what at one level may
be perceived as 'national' changes, may vary greatly in operation and in impact
between different parts of the country (Massey, 1984, 194). It is therefore crucial
not to simply accept national data without considering the possibility of spatial
variation through a breakdown of the figures at more regional and local scales. For
as Green et al. point out,

> 'Geographers and regional economists have long realised that the
> national economy is merely the aggregate of a set of regional and sub-
> regional economies. As such, national economic cycles will conceal the
> existence of distinct regional cycles.'
>
> (1994, 144)

This confirms the argument made in chapter 2, where I noted how academics had
identified the weakness of formulating policy which failed to acknowledge the
influence and importance of place specific processes.

It should be noted that the understanding that women's participation within
the labour market has significant regional and local variations, is not a recent
phenomenon. It was a perception expressed by the Women and Geography Study
Group (WGSG) in their first book in 1984, as they explored the establishment of
specifically female labour markets, following the decline in male jobs. They stated
that:

> 'There are additional inter- and intra-regional aspects to this pattern of
> social change. Consequently a district geography of women's
> employment has emerged since the war.'
>
> (WGSG 1984, 71)

The WGSG recognised that initial changes in female participation rates were
evident in the old industrial regions where male employment in such industries as

steel, shipbuilding and coal-mining began to decline and female labour market participation sharply increased. However, they also identified that the distinct regional variations were not solely a result of the decline in male employment, but also the dramatic proactive increase of women entering the labour market for the first time. The WGSG categorised and mapped regional employment change for both men and women in the UK into four distinctive economic regions; growth regions, intermediate regions, declining regions and the southeast (see figure 4.3).

Figure 4.3   WGSG map of regional employment change 1966-1978

Source: Reprinted with permission from WGSG 1984, p74

According to figure 4.3 the whole of Wales is an 'intermediate' region where total employment change is close to the national average because an above-average loss in male employment has been set off by an above-average gain in work for women. Although this may well be the case overall, there exists a wide variation within this one region. In the current economic climate some may argue that Wales could be regionally divided into two economic regions; south Wales, an economically developing and prosperous area, and mid and north Wales, whose main employer is the agricultural industry. According to the LFS spring 2000 figures, women's labour market participation within Wales ranges from 49 per cent in Neath Port Talbot to 77 per cent in Powys. Figure 4.4 demonstrates how the WGSG category of 'intermediate region' conceals widespread local variation of women's participation in the labour market. Similarly the southeast region, where male and female employment changes differ little from the national average, incorporates the areas of inner London which are heavily deprived with high unemployment. Variation within the London area alone ranges from 40 per cent in Newham to 76 per cent in Havering. These local variations in employment patterns are concealed by the categories which the WGSG use. However, the WGSG acknowledge that after further and more local analysis, other patterns did emerge for changes in female employment. They argued that the generalised patterns of employment change disguised the locational concentration of women's work within particular local areas. They also argued that the gender recomposition of employment involved a geographical reorganisation of local labour markets (1984, 73).

Figure 4.4   Variation of women's labour market participation in Wales*

Local authority district	% of women economically active
Bridgend	74
Caerphilly	54
Cardiff	66
Carmarthenshire	69
Flintshire	76
Neath Port Talbot	49
Newport	70
Powys	77
Rhondda, Cynon, Taff	59
Swansea	69
Vale of Glamorgan, The	61

Source: Labour Force Survey, Spring 2000

---

* Some counties are missing as they have too small samples to provide reliable data.

Another attempt to classify processes operational at the local level is that made by Cooke (1983b). In an analysis of labour market theory, Cooke feels that

'The lack of a spatial dimension in the analysis is surprising, especially in those theories which postulate segregation between labour markets, since *a priori* it seems reasonable to expect that space would exert some such effect upon the relationship between sub-markets.'

(1983b, 217-218)

In an attempt to classify the multiple processes which give direction to the dynamics of the local labour market, Cooke summarises three attempts by labour market theorists to add a spatial dimension to labour market theory. The first two approaches are the neo-dualist thesis (summarised by Berger & Piore 1980) and the core-periphery approach (led by Friedman 1977): both postulate that firms exist in a symbiotic relationship of independence. However Cooke argues that both approaches, while providing valuable explanations in the role of struggle in spatial development, tend to be over-constrained by the use of unnecessarily rationalistic conceptual frameworks (1983b, 221). The third approach attempts to derive explanations of spatially uneven development from a class-based analysis of variation between local labour markets (see Urry 1981, Massey 1980 and Cooke 1981). According to Cooke, this approach comes to the conclusion that labour markets cannot be satisfactorily conceptualised in rationalistic and functionalist terms such as 'core' and 'periphery'. Rather they are understood to be undergoing a permanent, dynamic process of recomposition or reconstitution in relation to the developing antagonism between accumulation imperatives and worker resistance (p218). Based on studies by Kreckel (1980), Urry (1981) and Cooke (1981), Cooke offers a breakdown of the various types of labour markets (e.g. underclass, precarious, feminised), the type of employee which that labour market attracts (e.g. unemployed, deskilled citizens, part-time workers and predominantly female), and where those firms are likely to be found (e.g. regional metropolitan and inner city, rural, and regional metropolitan respectively) (see figure 4.5). Cooke creates these typologies of socially differentiated labour markets in an attempt to explain the impact of spatial variation. Cooke gives an indication of the kind of jobs which may be expected to be found in various types of area, an explanation for this, and the extent to which the class interests of employees in such areas can converge to resist the recomposition effect of contemporary capitalism (1983b, 222-224).

Figure 4.5 A classification for spatially discontinuous labour markets

Predominant labour market	Main determinants	Primary class interest	Predominant location
Marginalised Illegal immigrants, criminalised, male and female non-citizens	Demand for 'sweated' labour, growth of informal or 'black' economy	Latent proletarian due to only having labour power to sell, but non-solidaristic	Regional-metropolitan and primate city ethnic enclaves, 'inner city'
Underclass Unemployed, redundant or deskilled citizens, labour reserves	Technological restructuring, 'runaway industries', excessive wage costs, anti-union closures	Latent proletarian because labour power not in demand, basis for solidarism fragmented	State-assisted areas, regional-metropolitan and primate inner-city ethnic and indigenous enclaves or 'traditional communities'
Precarious Guestworkers, part-time workers, limited contract and seasonal workers, male and female	Fluctuating demand for labour in unstable occupations. Inadequate non-work sources of income	Latent proletarian because labour power insufficiently in demand to ensure reproduction. Excluded from solidaristic organisations	Rural and various urban centres of labour-intensive employment, males in agricultural and construction work, females in miscellaneous services
Selective Temporary workers, periodic workers, male and female	Constant demand in tight labour markets, aggravated by employment protection law	Proletarian because only have labour power to sell, are not employers, and must be unionised	Regional-metropolitan and primate service centres, female secretarial/commercial, male/female periodic teachers, legal assistants, clerical
Feminised Full-time, service sector workers, predominantly female	Growth of state and private service sectors, routinised secretarial, data processing etc. work	Proletarian because only have labour power to sell, and likely to be unionised	Regional metropolitan (and some smaller) administrative centres, with substantial retailing and commercial functions
'Normal' compliant Small independent businesses, subcontractors, mainly skilled, male employed	Components suppliers to corporate enterprise, objects of 'productive decentralisation', competitive, low-wage, under-unionised	Latent proletarian due to anti-unionism of small employers, and advantages gained thereby for large firms	Periphery of industrial conurbations, semi-rural areas, possibly near branch plants
'Normal' resistant Large manufacturing firms, semi-skilled male employment, substantial female minority, especially in branch plants	Economies of scale, industrial monopoly, advanced mechanisation, strong internal labour markets, training, benefits	Proletarian because only have labour power to sell and unionisation vital to worker security	Industrial conurbations, branch plants in and near to assisted areas
Crafts (devaluing) Skilled manual employment, mainly in large manufacturing firms, predominantly male	Scarcity, monopoly of expertise, exclusivity, or if deskilling is occurring, inclusivity	Proletarian because only have power to sell, high solidarity, unionisation, demarcation	Industrial conurbations, centres of metal manufacturing, mechanical and electrical engineering

Crafts (revaluing)			
Skilled manual employment, specialised manufacturing and services. Small and medium-sized firms. Male and female	Scarcity, new skills or those which have been enhanced by technical change	Weakly proletarian because of potential for self-employment, solidaristic, exclusive, trade unionism	Mainly periphery of primate or large industrial cities, e.g. airports, power stations, R & D offices
Self-employed			
Small to medium-sized independent businesses, subcontractors, suppliers, primary extractive producers	Traditional petty-bourgeoisie, past and emergent demand for supplies to large firms	Capitalist due to purchasing labour power and producing surplus value	Periphery of industrial conurbations, semi-rural areas, suburbia, specialist inner-city 'quarters'
Subordinate functionaries			
Lower-order state and private sector administration, management and professional occupations	Growth of subordinate technical and administrative functions as complexity of division of labour increases	Capitalist where production is of or for surplus value. Proletarian where labour power is solely to meet social needs	Regional-metropolitan centres, primate city centres, local administrative centres
Independent functionaries			
Higher-order state and private sector administration, management and professional occupations	Increased separation of conception and decision taking in production, financing and administration of national and international capital state functions	Capitalist because main interest is in reproducing national and international conditions for surplus value production	Primate region or specialised centres of government, finance or production

Source: Reprinted with permission from Cooke 1983b, p226

Although Cooke provides an interesting analysis of typologies created through the recognition of the importance of spatial variation, this research does so mainly on the basis of economic derivatives governed by the labour market, with little consideration given to the interlocking social systems which also affect women's decision-making processes and subsequently the constitution of the local labour market such as household interdependencies, childcare and patriarchal expectations.

Subsequent research has built on these suggestions by Cooke and further evidence of the significance of regional spatial variation can be found in the work of Green and Owen (1998). Green and Owen explore the inter- and intra-urban variations in non-employment, and report the various pictures presented at national, regional and local authority district and ward level. They indicate that the variations differ according to the scale of analysis and assert that a reliance solely on conventional unemployment statistics can disguise the true picture of joblessness – to different extents in different areas (1998, 58). Their overall analysis saw an increase in non-employment levels in the decade from 1981. However when broken down by gender, 20 of the 459 local authority districts (LADs) saw a decrease in the non-employment rate for women of working age. These areas were mainly London boroughs or large cities in northern Britain (e.g. Manchester, Liverpool and Glasgow). In an attempt to present an explanation for the variation, Green and Owen use the ONS classification of local authorities (see Wallace and Denham 1995), and categorise the LADs into six 'families'; Inner London, Mining and Industrial areas, Maturer areas, Urban centres, Rural areas and Prospering areas. Once broken down into these categories, Green and Owen discovered that there was an increase in the non-employment rate for women only in the 'Inner London' area (as discussed in the next chapter, the unique social and economic characteristics of the Inner London area, may begin to provide an explanation for this).

In order to identify what the categories used by Green and Owen actually tell us about female labour market participation in differing economic conditions, I have divided the local authorities in my Labour Force Survey data into the economic categories provided by ONS and plotted them geographically. Figures 4.6, 4.7 and 4.8 demonstrate the variation of women's labour market participation using data at the local authority district level within three randomly selected ONS defined families; prospering areas, mining and industrial areas and rural areas.

Gender, Place and the Labour Market

Figure 4.6   Variation of women's employment in prospering areas

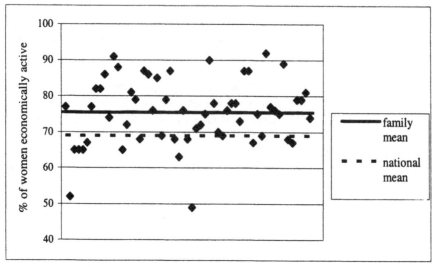

Source: *Labour Force Survey*, Spring 2000

Figure 4.7   Variation of women's employment in mining and industrial areas

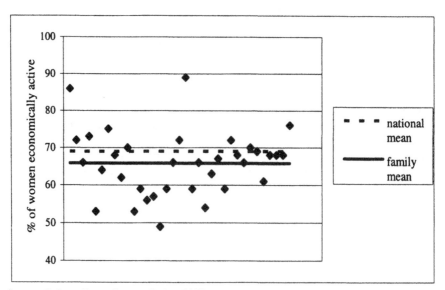

Source: *Labour Force Survey*, Spring 2000

Figure 4.8   Variation of women's employment in rural areas[1]

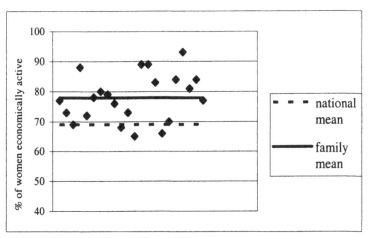

Source: Labour Force Survey, Spring 2000

According to the ONS classification of local authority district areas, prospering areas demonstrate well-above-average values in contemporary indicators of prosperity and good housing. They have low unemployment, high availability of cars and proportions of the highly qualified and social classes one and two. Prospering areas have the highest levels of owner occupation and large dwellings and the lowest levels of housing without central heating. When plotting my data onto the ONS classification, this family demonstrates variation in women's labour market participation from 63 per cent in Maidstone to 91 per cent in Cherwell. Similarly, mining and industrial areas correspond closely with the coalfields of Great Britain and associated primary production, heavy industry and services, often in coastal positions. According to the classification system, this family contains some of the most deprived parts of Great Britain, including the highest number of social classes four and five, second highest levels of unemployment and children with lone parents, and half of the people in this family are in households without cars. When we look more closely at the local authorities that are included in this family, again it demonstrates a large variation in women's labour market participation, varying from 49 per cent in Neath Port Talbot to 89 per cent in Renfew. Finally within rural areas, which are characterised by high agricultural employment, women's labour market participation varies greatly from 65 per cent in Nuneaton and Bedworth to 93 per cent in West Dorset. The reworked classifications do reveal different rates of female participation in the labour market. Rural areas and prospering areas are above the national mean while women's employment is below the national mean

---

[1] Although there are many more local authority areas which are classified as 'rural', in the calculation for the proportion who are economically active, often the figure was too small to be reliably calculated and could therefore not be included.

in mining and industrial areas. However the diagrams also reveal that this is not the whole story and the differences in economic type cannot be solely accountable for different participation rates. For instance, even though the overall participation rate is low in mining and industrial areas, some local authorities in this cluster have rates far higher than some of those in prospering and rural areas. Likewise, there are different numbers of local authorities within the prospering category whose female participation rates are below even the mean for mining and industrial areas. What this indicates is that perhaps there are other influences at work, in addition to the economic, helping to determine whether women work. In their analysis, Green and Owen conclude that 'the national picture disguises important variations in experience at the regional level' (p15). This might be the case, but after closer examination of the variation of women's labour market participation within each of the 'regional families', it must be questioned what Green and Owen's typology actually tells us about the geography of women's participation in the labour market.

Massey and Meegan (1982) argue that in searching for an explanation for the geography of job loss, it is not enough to rely on the economic characteristics of an area or a particular location. It is also necessary to understand the reasons why the processes that are causing the job loss are occurring. Massey and Megan's research demonstrates

> '...that the identification and explanation of the different mechanisms can also make a significant contribution to the geography of employment decline. All too often, attempts to explain the geography of employment decline – why some parts of the country lose more jobs than others – have concentrated on trying to find an association between the map of employment decline and the characteristics of different areas. We argue that this is not enough. First, it is analytically inadequate; the form of production change through which jobs are lost may be at least as important for the understanding of the geographical pattern as the locational characteristics may vary in the way they operate, depending on the kind of production change. Second, the difference between the two approaches is politically important. The approach which looks immediately for a correlation with the characteristics of declining regions tends more easily to slip into an argument which essentially 'blames the victim'. What we argue here is, not that such geographical characteristics are unimportant, but that their degree of importance, and the way in which they operate, depend on the kinds of production change that are causing job loss. It is not enough to go straight from the geography of decline to seek its explanation in geographical characteristics (e.g. Merseysiders go on strike more often). It is necessary to understand the reasons why, and the processes by which, the jobs are being lost in the first place.'
>
> (1982, 123-124)

Massey and Meegan go on to look at three particular processes which can lead to particular job loss; rationalisation, intensification and technical change, and acknowledge that each process will result in a different spatial variation, as particular locations are more likely to be affected by one of the processes than the others. In relation to an explanation for the variation of women's participation in the labour market, Massey and Meegan's investigation adds to the argument that it is not possible to construct typologies based solely on economic factors. It is necessary to understand the reasons why some women seek to combine employment and motherhood, whilst others wish to concentrate fully on one or the other, and see the interjection of the other as a barrier to enabling them to do their chosen task satisfactorily. The remainder of this chapter argues the necessity of gaining an understanding of important social factors which, in addition to economic factors, also influence women's decision on whether or not to enter the formal labour market. I shall therefore argue that it is a combination of both the economic and social which constitutes women's participation in the local labour market.

Locality as a Process: Constituting Local Labour Markets?

Many academics have argued that women's participation on the labour market is *more* than a reflection of the status of the local economy; it is in fact women's expectations, and the expectations of women, which are more influential in affecting the process of women entering the labour market (see Duncan 1989, 2000). According to Duncan and Smith (2002), regional and local differences in women's employment patterns are not only influenced by geographical variations in the availability of jobs at the local labour market level, or by the varying provision of welfare services supporting working women (such as those discussed in chapter 2). Instead, variations in women's employment rates are crucially influenced by local social and institutional ideologies and the perceived gendered roles about what men and women ought to do, and about their interdependencies. Duncan and Smith describe how in some places mothers are generally seen in the stereotypical homemaker role; that of caring for children at home, maybe with some part-time paid work when the children are at school. In other places mothers are primarily seen as workers and children are cared for by others during work hours. An understanding of how and why social and institutional ideas develop, and the extent of the impact caused by them in the decision-making process for women, is therefore crucial to explaining the variation of women's participation in the workforce.

The local social and institutional ideologies of what women's role should be are referred to by Forsberg as 'gender contracts' (1998, 193). According to him, these display 'regional' variations, and affect the daily life and the operating spheres of women and men in different ways in different regions. In support of this argument, Duncan and Edwards (1999) discovered that different social and ethnic groups of lone mothers in Britain showed different 'gendered moral rationalities' towards combining paid work and motherhood, demonstrating the influence of additional considerations other than the presence of formal support systems. Pfau-Effinger (1998) also argues that gender cultures, rather than differences in national social policy such as childcare provision, provide the primary explanation for national variations in women's paid work. She argues that such government initiatives as New Labour's National Childcare Strategy are a response to current demands from the female workforce who are struggling to combine formal paid work with being a full-time mother, rather than a breakthrough to provide alternative 'choices' for mothers. Similarly, Duncan and Smith describe how the economic and income effects of training and educational policies will partly depend upon whether mothers are expected – and expect – to be basically carers at home (perhaps with some part-time work organised around this prime responsibility) or full-time workers using childcare services (2002, 21). Duncan and Smith highlight historical evidence which shows that women entered the workforce in large numbers before the development of state support (2002, 22) (also see WGSG 1984), and argue, therefore, that local expectations of women's roles play a more important part in women's decision-making process. Duncan and Smith point out that this turns the normal social policy argument on its head, which assumes that it is variations in national welfare and tax/benefit policies which determine local differences in women's behaviour in combining paid work and motherhood. Any local differences then become, by implication, trivial and almost random variations around the national standard (2002, 22).

The movement of women into the labour market is not always accepted as a positive move. The WGSG argue that there is hostility expressed towards women's employment which stems from the threat it poses to the traditional image of male and female roles (1984, 75), that of man as breadwinner and woman as homemaker. This supports Pfau-Effinger's argument that both variations in women's uptake of paid work, and the different availability of welfare services associated with these (for example public childcare), result from differences in deep-seated and long-lasting gender cultures about the position of women in families and how they should be combining caring and economic work (Duncan and Smith 2002, 23). I will investigate how much impact the gender expectation has on women's decision about whether or not to enter into paid employment. This is only possible through research at the local level, which I shall turn to in chapters 6 and 7.

Having argued the need for an exploration of women's participation in the labour market at the local level, one of the biggest problems faced by researchers, is defining what 'the local' is. There have been many attempts to define the local labour market area, all of which have started with an expression of the difficulties of such a task (see Green and Owen 1990, Peck 1989b). However, Bradley argues that all local labour market definitions share one common feature, which is an attempt to specify boundaries between 'functional areas', according to economic activity (Bradley 1984, 67). As Massey argues, geographers have long experienced the problem of defining regions, and the question of 'definition' has almost always been reduced to the issue of drawing lines around a place (1994,

152). Massey continues to describe how 'the local' is different for each individual who attempts to define it; it is relative to their personal networks, relations, experiences and understandings, 'whether that be a street, or a region or even a continent' (Massey 1994, 154). These issues have also been discussed by academics trying to define a locality. In this section of the chapter, I shall explore the emergence of the locality debate and the subsequent appreciation of locality as a process. However, I question whether an exploration of local labour market processes probes far enough to provide an adequate understanding of labour market participation, asking whether we need to be examining home-work interdependencies, through an understanding of individual women's social, political and economic networks?

During the mid 1980s, after Britain had experienced major economic restructuring, the Economic and Social Research Council (ESRC) funded the establishment of three locality research programmes; The Changing Urban and Regional System (CURS), the Social Change and Economic Life (SCEL) programme and the Economic Restructuring, Social Change and the Locality programme. All three programmes emphasised the importance of spatial variations in the restructuring process by acknowledging that not only did space make a difference to how social processes worked, but that spatial variations were key to understanding social processes. In order to investigate further and develop the academic debate on locality, the programmes understood that a 'proper' understanding would have to be a spatial one, focusing, at least in part, on small-scale, sub-national 'localities' (Duncan and Savage 1991, 155). The ESRC programmes aimed to take the academic locality debate forward by establishing a link between theory and empirical research at the local level. Although these projects made a valuable contribution to the locality debate, in hindsight it should be noted that the focus of the research was on the economic restructuring, with less emphasis given to the social aspects of locality.

Duncan, in his attempt to define 'locality', argues that 'the concept of locality clearly rests on the idea that space makes a difference' (1989, 224) as it influences the ways in which social relations operate. Duncan argues that there are three ways in which 'space makes a difference'. Firstly, there are spatial contingency effects, where processes are constituted in particular places, which are influenced by the pre-existing nature of those places and other processes taking place at that time. For as Sayer argues, 'the relationship between causal powers or mechanisms and their effects is not properly fixed, but contingent' (1992, 107). Secondly, there are local causal processes or influencing factors. Duncan explains that while it might affect how processes work, spatial contingency cannot create causal processes, which may be derived locally. Finally Duncan believes that space matters to locality effects, for the contextual effects of local causal powers and spatial contingency may be so significant as to alter the nature of social structures in a particular place and hence alter social action (1989, 246-7). An understanding of the processes behind the constitution of the local labour market is therefore crucial for Duncan (2000), who believes that people are not only aware of their social and spatial contexts, but will act partly on that basis. In relation to women's labour market participation, this suggests that some women may (re)enter, or not (re)enter, the labour market as a result of local social and cultural factors such as expectation dictating what they should do. Duncan argues that it is not only the proportion of women in employment within a local labour market which can vary, but the consequential impacts of such activity. This may lead to a subsequent change in the character of the local area, and the expectations of women.

> 'The economic and social geographies created and influenced by spatial divisions of labour are also gender geographies, therefore, where women's roles, possibilities and expectations will vary.'
>
> (Duncan 2000, 17)

Duncan describes how some researchers fail to recognise and acknowledge spatial variation, or somehow treat spatial differences as 'less real than the norm' (p225), believing nationally aggregated data to be representative. In doing this, Duncan argues that those researchers are failing to appreciate the wide variation which may exist and therefore do not gain an accurate picture of local difference.

> 'The first problem of ignoring spatial variation is in some ways akin to the problem of the average which never exists in reality...The second problem of approaches that deny spatial differences – how can they account for all the detailed real world experiences that exist?'
>
> (Duncan 1989, 229)

Duncan (1989) raises the issue of whether the average, usually created through an aggregation of data, ever actually exists in reality. This is a question which needs to be addressed by government when forming national public policy. I shall address the issues surrounding the implications of spatial variation for public policy in chapter 8.

Duncan's second problem of accounting for the 'real world experiences' is addressed by Massey in her (re)discovery of Kilburn High Street. Massey argues that it is not the economy alone which determines our experience of space and place (1994, 148), but the connection of many interlocking systems of difference which creates an individual locality which consequently determines the way in which we interact with that place. Massey, therefore, claims that

> 'People's routes through the place, their favourite haunts within it, the connections they make, between here and the rest of the world, vary enormously.'
>
> (Massey 1994, 153)

This argument supports the notion that all localities are individual and are created through the connections of their many links with other places. It is therefore necessary to gain an understanding of local social, cultural and economic systems in order to appreciate and account for local variation from the overall generalised national aggregate. In Duncan and Massey's work on 'locality', it becomes clear that both of them find it difficult to accept any investigation of national trends, which has not at least acknowledged the importance of spatial variation as justifiable.

Conclusion

This chapter has explored the current level of variation in women's labour market participation within the UK. It has demonstrated that the variation in women's labour market participation is not solely a reflection of existing local economic trajectories and therefore cannot be categorised by economic typologies alone. Social and cultural factors are important influences producing local contingency 'effects' and 'local causal processes', which may in turn influence the ways in which local labour markets behave. Traditional images of male breadwinner, female homemaker may be reinforced through local patriarchal ways, or challenged through images that women can 'have it all' by being a valued member of the new economy as well as fulfilling their parenting role. I shall now turn to an empirical exploration of the geography of women's participation in the labour market.

Chapter 5

Research Design and Methodology: Moving Beyond the National

Introduction

Having argued for the necessity of local and meso level study in addition to national scale analysis, this chapter discusses methodological considerations in relation to gaining a further understanding of the complex networks within which women operate in negotiating their position between home and work. Throughout the research I have shared the view of Beechey, who argues that

> 'Workplaces are generally thought of as strictly demarcated from the home and family life. Similarly, we think of workers as people who leave home in the morning to travel to work and who work a certain number of hours...These conceptions of work and workers would be seen to be an accurate characterisation of some kinds of paid work. However, they are not very satisfactory as a characterisation of women's paid work.'
>
> (1986, 77-78)

This is due to the fact that women's ability to work, and the type of work they do, is often dictated by the responsibilities they have and undertake within their home and family life, and as a consequence these are often at the forefront of a woman's mind even when active in the workplace. In order to fully appreciate women's decision-making process with regard to entering the formal labour market, it is therefore necessary to explore the interaction and motivations between home and work for individual women, in addition to exploring the national level data and government policies which I set out in chapter three.

During the time of researching for this book, I was fortunate to undertake two work placements in the Women's Unit (WU) and the Women and Equality Unit (WEU) of the Cabinet Office.[1] This is a unique unit as it has its own minister, the Minister for Women, as well as close associations with other Cabinet Ministers. During my time in the WEU, I worked with policy advisors on issues relating to some of the dilemmas women face on a daily basis, such as childcare and work:life

---

[1] Originally called the Women's Unit, the unit took on extra responsibilities and consequently changed its title to the Women and Equality Unity. For ease of reference I shall refer to it solely as WEU from now on. As of 29th May 2002, the WEU moved from the Cabinet Office to be part of the Department of Trade and Industry (DTI).

balance issues. This was important to my research for it enabled me to see 'first-hand' how the government formulate policy and how it deals with academic and NGO research. I worked on two very different projects, not necessarily directly associated with my own research, yet overlaps were evident and the experience was invaluable, as I was able to gain a direct link to circuits of knowledge and view documents not accessible 'on the outside'. More specifically, I was able to identify first-hand what government priorities were with regard to women and labour market issues. Acknowledging debates in human geography regarding 'situated knowledge' (see Rose 1997, Hoggart et al. 2002, Merrifield 1995, Schoenberger 1992, Steaheli and Lawson 1994), my position as a researcher was enhanced through my increased knowledge of the way in which government saw the position of women within the labour market, and of its current policy in this area. As a researcher exploring issues surrounding parenting and labour market choices, a 'situation' I had no first-hand experience of, this also gave me more confidence in engaging with my research participants. I gained access to information, which not only enlightened my knowledge of the ways in which policy is formed (and informed), but also of the types of research and data that government departments work with, and the types they disregard. Often unless research is commissioned directly by themselves and/or another government department, it is disregarded as secondary and not reliable, although it may well form the basis of future research to be commissioned.

My experience gained from working in central government reaffirmed the importance of national scale data, for all the data which the government works with is on a national scale, with very little sensitivity given to spatial variation. The only time at which local difference was engaged with at all was when finding examples of 'best practice' to support the policies currently in discussion, or to have some local statistics available to the Minister for Women, if a local visit was being made. My experience of working directly with policy advisors in the Cabinet Office sharpened my opinion as a geographer that we should move beyond the national level and enquire about the spatial variation below this, in order to appreciate the impact an understanding of local variation might have on the formation of national policy.

The book has, thus far, concentrated on extensive data through an exploration of national policy, which is based upon nationally aggregated data as presented in chapter 3. However, chapter 4 discovered that in fact nationally aggregated data tells us very little about women's ability to enter into the formal labour market, by demonstrating the spatial variation in women's labour market activity across the UK. In this chapter, I shall firstly argue that, although extensive research techniques are important to uncover national labour market geographies, they are limited in their ability to provide explanation and reasoning for the locally articulated nature of employment. I shall argue that intensive research techniques can be used in conjunction with extensive methodology to provide a further depth of knowledge. Secondly, I shall discuss the use of case studies as a way of approaching intensive research and set out how I came to choose the two case study areas. I shall describe the research techniques I used to gain access to mothers and discuss with them the complex networks in which they undertake their daily lives. Finally I shall introduce the two case study areas, providing an economic context of the areas, informed by an extensive survey I did with several hundred women in the two areas.

Extensive and Intensive Research

The research for this book has involved a combination of both extensive and intensive research techniques. The former, using national data analysis and extensive questionnaires, has allowed me to gain a broad understanding of women's labour market activity. The latter based on qualitative research techniques such as semi-structured interviewing has enabled me to explore individual women's lives, choices and their decision-making process. The two types of research, extensive and intensive, have different objectives and use different techniques and methods to ask different sorts of questions (see figure 5.1). Sayer describes how, in intensive research, the primary questions concern how some causal process works out in a particular case or limited number of cases. Extensive research, which is more common, is concerned with discovering some of the common properties and general patterns of a population as a whole (1992, 242). This chapter will demonstrate how intensive research techniques can be used alongside extensive techniques to gain a full understanding of women's labour market activity.

Figure 5.1   Intensive and extensive research: a summary

	Intensive	Extensive
Research question	How does a process work in a particular case study or small number of case studies? What produces a certain change? What did the agents actually do?	What are the regularities, common patterns, disguising features of a population? How widely are certain characteristics or processes distributed or represented?
Relations	Substantial relations of connection	Formal relations of similarity
Type of groups studied	Causal groups	Taxonomic groups
Type of account produced	Causal explanation of the production of certain objects or events, though not necessarily representative ones	Descriptive 'representative' generalisations, lacking in explanatory penetration
Typical methods	Study of individual agents in their causal contexts, interactive interviews, ethnography. Qualitative analysis	Large-scale survey of population or representative sample, formal questionnaires, standardised interviews. Statistical analysis
Limitations	Actual concrete patterns and contingent relations are unlikely to be 'representative', 'average' or generalisable. Necessary relations discovered will exist wherever their relata are present, e.g. causal powers of objects are generalisable to other contexts as they are necessary features of these objects	Although representative of a whole population, they are unlikely to be generalisable to other populations at different times and places. Problem of ecological fallacy in making inferences about individuals. Limited explanatory power
Appropriate tests	Corroboration	Replication

Source: Sayer 1992, 243, reprinted with permission

Although figure 5.1 demonstrates that intensive and extensive research methodologies are very different in their objectives, they can also be used together.

It is often said that extensive methodologies produce results that are 'representative' of the sample population, and that generalisations cannot be made from research based on individual case studies (see Mitchell 1983). For this reason, policy makers may favour quantitative research. Tangri and Strasburg (1979) have analysed the problems of utilising women's research for policy formulation and offer some recommendations for generating relevant research. Tangri and Strasburg specifically mention that the researcher should be perceived by the policy-maker as 'objective', and produce findings, which are statistically 'significant', therefore, demanding the use of extensive statistical data analysis. Extensive research data can be found in chapters 3 and 4 where I have displayed the national picture as presented by government and academics in their broad assessment of women's participation in the labour market. Although perhaps necessary in order to compile policy to 'help' women have choices (see chapter 3), such extensive data, with its picture of a UK nationally aggregated figure of 69 per cent of women economically active, does not shed much light on what is happening in local areas (see chapter 4).

Extensive research methodologies normally generate quantitative data which can be used in descriptive or inferential statistics and other forms of numerical analysis. Often this comes from a large-scale formal questionnaire of a population or representative sample. In extensive studies, Sayer explains how the criteria by which samples are drawn are decided in advance and adhered to consistently in order to ensure representativeness (1992, 244). Many authors agree on the need for research to be as objective as possible (see Duncan and Duncan 2001, Epstein Jayaratne 1997), even though accepting that personal biases impinge on the research process in many ways; through theory formulation and interpretation, development of design, data collection and analysis. However, quantitative data analysis can be representative of a whole population and is specifically developed to produce an objective evaluation with little room for personal interpretation and is therefore less subject to biases, which can be present in qualitative research.

In contrast, intensive research is concerned with generating meaning and understanding, and allows for a less formalised, less standardised and more interactive research process in which claims are made that the researcher has a much better chance of learning from the respondent what the different significances of circumstances are for them (Sayer 1992). In comparison to extensive research techniques, Sayer describes intensive techniques as mainly qualitative methods such as structural and causal analysis, participant observation and informal interactive interviews (1992, 221). Even the most complex and sophisticated quantitative research cannot import the same 'in-depth' understanding of respondents as, for example, a thorough case study. The advantage of the structured questionnaire is that it provides comparability between regions among some key demographic and socio-economic characteristics of a large population. However, the disadvantage of this approach is that it masks individual variety and may attribute a consistency and stability to opinions which are rarely found in everyday life. As Durant et al. put it,

> 'Surveys necessarily impose a rigid observation 'grid'…as a result they detect some things and miss others.'
>
> (2000, 133)

Having highlighted the focus and limitations of both intensive and extensive research methodologies, I feel there is a strong argument for using these techniques alongside each other. As previously mentioned, extensive methodologies strive to be representative of the population. However, Sayer questions extensive methodologies as 'representative of what?' (1992, 249). Often a statistical average is found, gained from aggregated data at an extensive scale, to which no real individuals correspond (this point was further argued in chapter 4). However,

> 'If the population is not too diverse, it may be possible to define taxonomic classes in which individuals share similar causal powers and liabilities, hence enabling extensive and intensive research techniques to become more complementary. Both methods are needed in concrete research although the latter tend to be undervalued.'
>
> (Sayer 1993, 250)

In intensive studies, the individuals need not be typical or representative of the whole population for, as Sayer points out, providing that there is no pretence that the whole population is represented, there is no reason why an intensive study should be less objective than an extensive one (1992, 226). It is believed that such intensive methodological approaches draw on women's purported abilities to listen, to empathise and to validate personal experience as part of the research process (McDowell 1992). Sayer concludes with this point:

> '[If] much of the information is qualitative and concerns processes, activities, relations and episodes of events rather than statistics on particular characteristics…the interdependencies between activities and between characteristics can be revealed; for example how waged work and domestic work are integrated in time and space.'
>
> (Sayer 1992, 242)

Cook and Crang argue that qualitative, ethnographic methods, can be used to 'understand parts of the world as they are experienced and understood in the everyday lives of people who actually 'live them out' (1995, 4). In relation to research on women's negotiation between home and work, intensive methodologies allow for the unpacking of experiences, interdependencies and the process of decision making of women in an informal setting. In the words of Kent and Sandstrom, intensive research methodologies strive 'to interact with them in the most trusted way possible – without having any explicit authority role' (1998, 17).

By combining qualitative and quantitative research techniques I deployed a triangulation methodology, which according to Baxtor and Eyles is one of the most powerful techniques for strengthening creditability of research (1997, 514). They define method triangulation as the corroboration of constructs based on information derived from at least two methods. Triangulation suggests that a researcher can get a clearer picture of the social reality being studied by viewing it from several different perspectives. Combinations of qualitative methods are often used, however, and increasingly qualitative and quantitative methods are combined in the same study (see Harrison and Burgess 1994 and Leckie 1993).

Case Studies as a Way of Approaching Intensive Research

I obtained data from the LFS on the proportion of economically active women with dependent children[2] at the local authority district level.[3] [4] This data demonstrated the huge spatial variation that exists in women's participation in the UK formal labour market, as discussed in chapter 4. In order to find an explanation for the variation and to explore in-depth the complex processes and interdependencies which women negotiate on a daily basis, it is necessary to look closer within localities to find out why individual women had made particular choices regarding their working career (for varying explanations see Hakim 2000, Hanson and Pratt 1992, 1995, Joseph 1983, Johnson-Welch et al. 2000, Walby 1999). Schramm (1971) describes the case study as a way in which to try and illuminate a decision or set of decisions and to understand why they were taken, how they were implemented, and with what result. McDowell also believes that

'To me, this type of work – detailed case study investigations with what may seem like a small group of individuals – is a key way of exploring and understanding some of the processes behind the patterns of inequality and discrimination that mark and mar urban societies in industrial economies. The findings and conclusions from this work may not be equally applicable to all young men leaving school in all cities in Britain. Clearly the history of economic change and restructuring affects labour market opportunities. But it is only in the type of interactions made possible by personal engagements between a researcher and her/his subjects that we may come to understand why people in different places make the choices they do.'

(McDowell 2001b, 213)

---

[2] Dependent children are up to 16 years, or 18 years if they remain in full-time education.
[3] There are 409 local authority district in the UK.
[4] Data was based on figures collected during spring 2000.

It is precisely this understanding of the decision-making process, with regard to what had influenced women into making the 'choices' they had regarding their labour market activity, that I wanted to achieve with the intensive research.

Much time and effort can go into choosing case study areas. However, as Mitchell argues, 'there is absolutely no advantage in going to a great deal of trouble to find a 'typical' case...any set of events will serve the purpose of the analyst if the theoretical base is sufficiently well developed to enable the analyst to identify within these events the operation of the general principles incorporated into the theory' (1983, 204). In some circumstances a 'typical' case works well, however Stake (1995) points out that it is an unusual case which helps to illustrate matters we overlook in a typical case.

Mitchell highlights that probably the most often cited criticism of case study research is that of the invalidity of being able to generalise from one case study.

> 'Case studies of whatever form are a reliable and respectable procedure of social analysis and that much criticism of their reliability and validity has been based on a misconception of the basis upon which the analyst may justifiably extrapolate from an individual case study to the social process in general.'
>
> (Mitchell 1984, 207)

However, Stake (1995) argues that although only a single case, or a few cases may be studied, they will be studied in-depth and certain activities, problems and responses will come up repeatedly. Human behaviour, thoughts and feelings are partly determined by their context. According to Gillham (2000), if you want to understand people in real life, you have to study them in their context and in the way they operate. Gillham believes that the meticulous description of a case can have an impact greater than any other form of research report (2000, 101).

The case studies in this research were not chosen for their 'typicality', rather for their ability to help me explore the process behind women's decision making, regarding their decision to work or not work. I felt that the processes underlying women's decision making would be easier to reveal in extreme circumstances. Therefore in order to choose my case study areas, I ranked local authority districts areas with the highest and lowest women's labour market participation (see figure 4.2). Choosing a case study from each extreme would help me to investigate whether women in high labour market participation areas were *choosing* to work, and whether women in low labour market participation areas were *choosing* to stay at home. As figure 4.2 shows, many of the local authority districts with low participation rates are in the Greater London area. However I felt that this area was not the best to choose for a case study due to the unique characteristics associated with living and working in London. As Bennett and McCoshan point out,

'London is a special case which requires separate discussion. Although it exhibits many features of large cities and their fringes, particularly under-bounding, it has some key characteristics which place it in a unique position. First, London's core contains most of the largest national and international companies in the UK. Their networks are of national rather than local importance. Most major businesses, as a result, do not see a clear local responsibility to London itself (although there are some exceptions to this). Second, central London's businesses are dominated by financial services, tourism and retailing which are sectors that generally have had a lower tradition of involvement in local affairs. Third, there are enormous flows of workers from the suburbs into the central core, which undermines the source of employer-employee relationships, i.e. economic under-bounding is extreme. Fourth, and related to these factors, London constitutes a huge integrated labour market with few focal and identifiable sub-markets (Heathrow and Croydon are exceptions). Fifth, rapid economic growth in the rest of the South East has attracted an increasing set of flows of reversing commuting and economic linkage outward as well as inward. As a result, there is a high degree of economic interlinkage and flux into, across and outside London which is not found to the same extent in other metropolitan areas. This means that its problems require to be tackled on a London-wide or even South East scale, as well as in more local contexts. Given its size, this presents a formidable challenge.'

(1993, 225)[5]

Due to the difficulties of studying labour markets in London as described by Bennet and McCoshan and the desire to explore processes outside of the labour market, I decided to take the highest and lowest local authority districts which were not within a London borough as my case study areas; Neath Port Talbot, which has a female economic activity rate of 49 per cent, and West Dorset, with a female economic activity rate of 94 per cent (LFS spring 2000).

I shall now discuss the combination of extensive and intensive research used in order to learn more about women's labour market activity within the case study areas; structured questionnaires, in-depth interviews and documentary analysis.

Questionnaires

In order to gain more information about the labour market activity of the women in the two case study areas, I decided to distribute a structured questionnaire asking mothers some basic questions on their family status, employment status and their previous labour market activity. This enabled me to gain a more detailed knowledge of the labour market patterns in the areas as well as extensive data on the kinds of decisions women in the two areas had taken with regard to labour market participation. When considering the best way in which to access mothers, I decided to distribute the questionnaires through schools in the area.

---

[5] Reprinted with permission of Sage Publications Ltd from Bennett and McCoshan (1993) *Enterprise and Human Resource Development: local capacity building.*

This meant not only an immediate access to mothers, but I also felt that respondents might be more inclined to respond to a questionnaire sent home from school rather than a random letter drop or a stranger in the street, therefore potentially giving me a higher response rate. I received a total of 685 questionnaires from a variety of primary, junior and secondary schools across Neath Port Talbot and West Dorset; this was an overall response rate of 19% (19.9% in Neath Port Talbot and 20.5% in West Dorset). Although using schools to distribute the questionnaires provided many advantages, as discussed earlier, there are also issues to be cautious of when using schools as part of the research process. Firstly there may be social differences in the catchment areas of the schools which may affect the response rate, type of woman and the conclusions that can be drawn. However, my sample of schools covered a range of ages and types of school which are broadly representative of the area. Secondly, the headmaster and school administrator are acting as gatekeepers and their level of promotion for the survey may also influence the response rate. A final consideration relates to the school's distribution of the questionnaire where two siblings attend the school. Does each child get a questionnaire, or does each mother get a questionnaire? The results from these questionnaires will be discussed later in this chapter when I introduce the two case study areas.

Undertaking Intensive Research – Informal Interviews

After conducting a survey of mothers in the two case study areas, I felt it necessary to engage with individual women to discover more about the complex networks in which they operate. In-depth interviews with women would allow me to learn about the way in which women view their world and the ways in which they negotiate between mother and worker. For as Sayer (1992) argues, intensive research techniques allow for an explanation of the production of events through the studying of individual agents (see figure 5.1).

At this stage I feel it is necessary to briefly engage with arguments of feminist methodology. Does the fact that I am researching women call for a specifically feminist methodology? Hammersley argues that it is widely accepted amongst feminists that research on women implies a distinctive approach to enquiry, for 'feminism is taken to carry distinctive methodological and epistemological implications' (1995, 45). According to Du Bois, feminist methodology is about

> '...addressing women's lives and experience in their own terms, to create theory grounded in the actual experience and language of woman, is the actual agenda for feminist social science and scholarship...To see what is there, not what we've been taught is there, not even what we might wish to be there, but what is.'
>
> (Du Bois 1983, 108)

Du Bois is trying to emphasise that a researcher must listen to the personal experiences of their participants in order to create an actual account of how things are, rather than how they are perceived to be. If this is a definition of feminist research, then I would have to agree that this is what I have done. My interviews consisted of many hours listening to women's personal experiences with the intention of using their experiences to interpret labour market theory.

A common feature of feminist methodology is an emphasis on the validity of personal experience as against the conventional emphasis on scientific method, and structured statistical analysis. The feminist emphasis on the validity of direct experience is associated with the idea that women have uniquely valid insights, which Hammersley calls a 'feminist standpoint epistemology' (1995, 52). Finch believes where a woman researcher is interviewing another woman, 'this is a situation with special characteristics conducive to the easy flow of information' (1984, 74). Finch later explains that the 'special characteristic' comes from a shared subordinate structural position by virtue of their gender, which creates the possibility that a particular kind of identification might develop. This 'feminist standpoint epistomology' is the belief that only a researcher of the same identity can fully understand an ethnography of a participant for example, a woman can fully understand a woman, similarly only a black person can fully empathise with a black person. Reinharz emphasises that feminism revalues experience as part of social science methodology (1983, 167). When I consider my research, I am a woman interviewing a woman; however, all the women I met with were also mothers, something which I am not. Does this therefore mean that I will not be able to fully empathise and understand their situation?

Another issue for consideration when interviewing women comes from Beere (1979), who raises the idea of women faking responses, as they do not want to 'face reality'. Hyman et al. (1954) also discuss how participants will give different responses depending on the sex of the researcher. Beere describes faking as 'giving socially desirable responses rather than honest attributes' (1979, 385). This may happen consciously or unconsciously, and is not necessarily referring to information given just to researchers; it may be in discussions with friends or neighbours. Beere argues that it may be that faking is necessary for the psychological survival of many women because, without faking, reality would seem unbearable. Although I understand and recognise that Beere's argument may well be true in some cases and, as Gillham (2000) also acknowledges, there is a common discrepancy between what people say about themselves and what they actually do, I do not feel that it is appropriate in the case of my research. Discussions which I had with mothers were informal and non-judgemental. Respondents were not justifying their actions to me, rather they were informing me of the path they had taken, and stating what was important to them, and what factors influenced their decision regarding entering or not entering the labour market. Not being a mother myself and not having been forced to choose a particular path, the women I spent time listening to were not having to justify their actions, should they have done the opposite, to me. I therefore fell that this differentiation was a benefit to our relationship.

In a closer examination of the interview process Oakley (1981) refers to the traditional research technique of interviewer-interviewee hierarchy, displaying no emotional involvement, as a masculine mode of research and in opposition argues for a more interactional, less hierarchical, democratic, 'feminist' form of interviewing, in which the relationship of the interviewee is non-hierarchical and in which the interviewer is 'prepared to invest his or her own personal identity in the relationship (p41). Herod (1993) and Hyman et al. (1954) also present the benefits of woman-to-woman research, and support the rejection of hierarchy in a research relationship.

Although there are many aspects of my research methodology which engage with feminist ideas and practices, I also understand and appreciate the need for extensive methodologies often labelled masculine methodologies, which are desired by policy advisors and government institutions. It is essential to adapt research methodology to the context of the research and the type of information the researcher wishes to extract.   Therefore, although I was interviewing women, and a feminist methodology may be deemed appropriate for reasons explained earlier, I felt it was necessary to also engage with more 'masculine'-labelled research techniques to ensure my research retained validity in policy terms.

I completed 38 semi-structured interviews mainly conducted informally in the women's own homes; alternatively, for convenience, a couple of women preferred me to meet them at their place of work. Semi-structured interviews enabled me to gain an insight into the ways in which individual women negotiate their home and work interdependencies and to explore the influences on, and the reasons behind, the 'choices' they had made, whether to stay at home with their child or to continue with their formal employment. All interviews followed an informal conversation where I had set out the broad parameters for discussion, rather than a formal structured interview and lasted between thirty minutes and two hours. The interview was often led by the woman herself. Although there were some key questions and themes I wanted to discuss with each individual, often the conversation just flowed, following the life story of the woman from when she left education, understanding the choices which she had made as she hit key stages in her life (for example marriage, birth of child, divorce), through to her present day situation, and future intensions. Although I set out with a list of questions which I wanted to make sure that I covered with each woman, and had a different set for the women in work, and the women not in work, however, these were often reformulated during the course of the interview, as information provided by the women meant that some questions were no longer appropriate and new questions may have been necessary. The findings from these interviews will be discussed in chapters 6 and 7.

In addition to the distribution of a structured questionnaire and semi-structured interviews in the case study areas, I have also analysed local authority documents such as Economic Development Strategies for each area and the Early Years Development Plans to gain a depth of contextual knowledge of issues such as childcare, the main local employment industry and the status of the local economy. The next section of this chapter provides a summary of these findings, introducing Neath Port Talbot and West Dorset as my case study areas.

Introducing the Two Case Study Areas

Although the two case study areas were chosen based on the women's labour market activity, in this section I shall provide a broad economic overview of the two case study areas, incorporating results from my questionnaire to give a broader picture. This will provide a context within which women make their labour market decisions. Figure 5.2 provides an economic overview of the two areas, and shows that, although average weekly income in West Dorset is lower than in Neath Port Talbot, the economy appears more vibrant with higher employment and high levels of education qualifications gained. The reliance on the service sector rather than on manufacturing perhaps also gives West Dorset a more stable economic future.

Figure 5.2   Comparison of two case study areas

	Neath Port Talbot	West Dorset	UK national average
Area	442 sq. km	417 sq. km	N/A
Population	134,470[a]	92,350[a]	N/A
Employment rate	63[b]	79[b]	74[c]
Economic inactivity	26[b]	10[b]	21[c]
% of workforce unemployed	6[b]	Too small for reliable estimate[b]	3[c]
% of population over retirement age	18[a]	24[a]	18[a]
Average gross weekly earnings	£417.30[d]	£391.30[d]	£464.70[d]
Employment mix	Manufacturing and Service[b]	Service[b]	Service[b]
Number claiming Working Family Tax Credit	3281[c]	1864[c]	1,293,000[b]
% of population with NVQ level 3+	30.5[b]	48[b]	41[b]
% of working age population with no formal qualifications	24[b]	6[b]	15[e]

Key to sources
N/A   Not applicable
[a]   Census 2001
[b]   National Statistics Local Labour Force Survey (February 2002)
[c]   Labour Market Trends (February 2002)
[d]   2002 New Earnings Survey
[e]   DfES January 2003

Neath Port Talbot

Neath Port Talbot is an ex-mining and industrial area in South Wales, covering an area of 442 square kilometres, extending from the heavily industrialised coastal strip of the steel works and assorted petrochemical plants to the fringes of the Beacon Beacons National Park. Whilst the hinterland valleys were once a thriving coal-producing area, during the past three decades the economy of Neath Port Talbot has undergone major changes, which has resulted in almost the complete closure of coal-mining in the Welsh Valley area, along with major closures and rationalisation in steel and petrochemicals. Attracting new businesses is therefore a high priority and the area has benefited from the availability of European Structural funds. In the steel industry, although jobs have been lost through improvements in productivity, actual output has increased and the recent £120million investment in the Continuous Annealing Process Line represents the largest single investment in the British steel industry in the last decade.

In 1998, 3,500 job losses were announced, mainly in manufacturing. This was a trend that was set to continue, with competitive pressures to reduce labour costs and increase productivity within the manufacturing industry. Job loss announcements over the past three years have far outweighed job gains with a preponderance of losses in the well-paid manufacturing sector and a high proportion of gains in the lower-paid retail sector. Employment decline has resulted in a population decline, as people move away from the Neath Port Talbot area in search of employment (see figure 5.3). In March 2002, five per cent of the workforce were unemployed (NPT CBC 2002) with particular concern for those in the 16-24 age group which represented 32 per cent of the unemployment in Neath Port Talbot. In February 2001, the working age employment rate was 57 per cent compared with 74 per cent in the UK (LFS Local Authority Database) with only ten per cent of the population being self-employed (Wales Year Book 2002). The decline of well-paid jobs in manufacturing, including steel, chemicals, oil and engineering, coupled with the decline of coal, has resulted in a fall in the area's GDP from 99 per cent in the mid 1970s to only 87 per cent of the UK average in 1999 (NPT CBC 1999).

Figure 5.3   Population change in Neath Port Talbot

Year	Population (thousands)
1981	142.7
1991	139.4
2001	134.4

Source: National Assembly Wales 2001 and Census 2001

The economy today is diversified, with light industry replacing the diminishing heavy industrial sector. Much of the job creation in the area is provided by part-time jobs, particularly in the retail sector, although 32 per cent of total jobs in the area remain in manufacturing (see figure 5.4) compared with 19 per cent for Wales and 15 per cent in Great Britain (NPT CBC 2002). The average gross weekly earnings in Neath Port Talbot is the highest in Wales at £417.30 (New Earnings Survey 2002), reflecting the continued importance of well-paid manufacturing jobs to the local economy. However, the generally high rates of pay for those in employment disguises the high economic inactivity rate of 38 per cent, which compares with 26 per cent for Wales and 21 per cent for Great Britain (NPT CBC 2002).

Figure 5.4   % of jobs in each industrial sector

Sector	Neath Port Talbot	Wales	Great Britain
Primary	1.7	2	1.7
Manufacturing	31.7	18.6	15.1
Construction	8	5.3	4.5
Service	58.5	74.1	78.8
Total jobs	40,000	1,077,600	25,140,600

Source: Adapted from Neath Port Talbot Economic Development Strategy 2002

The questionnaires conducted with mothers in the area provided me with a further breakdown of women's labour market activity figures within each of the areas. When embarking upon my research into Neath Port Talbot, I had the information (from LFS data) that 49 per cent of women were economically active, yet as can be seen in figure 5.5, 73 per cent of my respondents were economically active, with the majority of respondents employed part-time at the point of filling in the questionnaire (February 2001). However, what should also be noted is that more respondents classified themselves as housewives (21) than employed full-time (19 per cent).

Figure 5.5   Employment status of respondents in Neath Port Talbot

Employment status	Neath Port Talbot %
Employed full-time	19
Employed part-time	46
Self-employed full-time	2
Self-employed part-time	2
Currently on maternity leave	1
Unemployed, actively seeking work	3
Total economically active	*73*
In full-time education	1
In part-time education	1
Retired	0
Unemployed, not actively seeking work	3
Housewife	21
Total economically inactive	*26*
Other	1

When seeking for an understanding of women's labour market activity, it is interesting to look at other factors such as marital status and, if applicable, the type of employment undertaken by the partner, for if the partner is in particularly demanding employment, the mother might be further restricted as to the type of employment she is able to undertake. The majority of my respondents were married (75 per cent) or cohabiting (10 per cent) at the time of filling in the questionnaire and indicated that 48 per cent of partners were employed in the construction and manufacturing industries. These are industries that often involve shift work and overtime; consequently the mother's employment or childcare options would have to be flexible enough to work around the partner's demanding hours. When asked to consider the importance of finance in their decision regarding whether or not to enter into the formal labour market, as can be seen from figure 5.6, only 3 per cent indicated that finance played no part at all in their decision.

Figure 5.6   The importance of finance in decision to enter or not enter the formal labour market

| | Neath Port Talbot |
	%
Not at all important	3
2	0
3	6
4	11
5	17
6	16
Very important	47

As previously discussed, the pressure of woman's dual role of mother and worker can strongly influence the type of employment she is able to undertake (see chapter 2). This is often due to the fact that the woman remains the primary carer for the children and her employment must fit around such events as school times. However, for Neath Port Talbot, the majority of women (62 per cent) indicated that having a child had not affected the type of job they did. As can been seen from figure 5.7, the majority of women returned to work after their baby had reached the age of two years with 61 per cent of mothers not returning to the same job they held prior to the birth.

Figure 5.7   Age of child when mother returned to work

| Age of child | Neath Port Talbot |
	%
0-3 months	16
3-6 months	12
6-12 months	9
12-18 months	5
18 months-2 years	3
2-5 years	34
5-10 years	17
10-16 years	4

West Dorset

West Dorset is a tourist area on the south coast of England, covering an area of 417 square kilometres, which includes five major towns and 25 miles of coastline. West Dorset is mainly rural in nature and has 49 square kilometres that are classified as Areas of Outstanding Natural Beauty. The West Dorset area currently has a lower population than Neath Port Talbot, but in contrast population has been steadily rising as can be seen in figure 5.8.

Figure 5.8   Population change in West Dorset

Year	Population (thousands)
1971	74.0
1981	78.3
1991	86.7
2001	91.6

Source: Adapted from Census and West Dorset District Council Performance Plan

West Dorset has a relatively affluent and very diverse economy, with the service sector being the main employer in the area, with 130,000 employees (WDDC's Performance Plan, 2000, 9). The majority of businesses are small employers, with 85 per cent having ten employees or less and 42 per cent with under five employees (WDDC 2000, 9). Tourism accounts for 25 per cent of the district's economy, receiving a steady 2.5 million visitors a year generating £121 million expenditure (WDDC 2000, 1). However, due to the rural nature of the district, the impact of structural changes in the agricultural sector has been, and will continue to be, significant on the local economy. The West Dorset District Council is trying to promote diversification for mainstream farm operators (WDDC 2000, 5) as a method of survival. The average full-time wage for men and women in West Dorset is £391.30 per week, one of the lowest in the South West. However, the West Dorset District Council Economic Development Strategy (WDDC EDS) recognises that anomalies such as a low average wage and pockets of deprivation in Bridport and Dorchester reveal a hidden need for regeneration activity (2000, 5). The WDDC Economic Development Strategy asserts that the survival of market towns is key to the prosperity of the West Dorset economy. However, the proportion of economically inactive in West Dorset is also relatively high as there is a trend for inward migration to the area for early retirement.

Similarly to Neath Port Talbot, the majority of respondents in West Dorset (48 per cent) were employed part-time at the time of filling in the questionnaire (February 2001), with 22 per cent employed full-time (see figure 5.9). However, in comparison with Neath Port Talbot, significantly[6] more women in West Dorset are self-employed both part-time and full-time, and significantly fewer women are unemployed, actively seeking work or housewives.

---

[6] Statistical analysis allows us to see which group differences are likely to be due to random variations between individuals. All group differences here are reported at a 5% level, which means that I can be 95% confident that they did not arise by chance.

Figure 5.9   Employment status of respondents in West Dorset

Employment status	West Dorset
	%
Employed full-time	22
Employed part-time	48
Self-employed full-time	7
Self-employed part-time	7
Currently on maternity leave	1
Unemployed, actively seeking work	0
Total economically active	*85*
In full-time education	0.5
In part-time education	0.5
Retired	0.5
Unemployed, not actively seeking work	0.5
Housewife	11
Total economically inactive	*13*
Other	2

The majority of respondents in West Dorset were split between ages 30-39 years (47 per cent) and 40-49 years (43 per cent), which is significantly older than respondents from Neath Port Talbot, indicating that women in West Dorset are having children later in life. This could be a reflection of the types of employment the women are in, with more women in West Dorset wanting to maintain careers, rather than occupying lower status retail and manual labour jobs. These figures reflect the fact that West Dorset is a more vibrant and buoyant economy. There is a positive significant correlation between whether the respondent's mothers worked or not and whether they entered the formal labour market. This was explored further in my in-depth interviews and shall be discussed in chapters 6 and 7.

Similarly to Neath Port Talbot, the majority of respondents in West Dorset were married (78 per cent), although some were cohabiting (8 per cent). However many respondents also indicated that their currently marital status was divorced (11 per cent), suggesting that at the time of filling in the questionnaire they were living as single parents. In an evaluation of partners' occupations the majority were employed in other services (25 per cent) and the construction industry (20 per cent). As can be seen in figure 5.10 finance played as much a part in the women's decision-making process regarding entering the labour market for women in West Dorset as it did in Neath Port Talbot.

Figure 5.10   The importance of finance in decision to enter or not to enter the formal labour market

	West Dorset %
Not at all important	4
2	4
3	4
4	12
5	18
6	13
Very important	45

Due to the more affluent economy, it may be expected that women in West Dorset would be returning to work sooner, as they may be in better paid careers and able to afford childcare, and that those in Neath Port Talbot were more likely to remain at home at least until the children went to school, as they are in less career-minded employment and unable to afford to pay anyone else to look after the child. This is demonstrated in figure 5.11. However, the same broad pattern of returning to work is evident, with the highest number of respondents in each case returning to work either before the child is six months old or between two and five years. The majority of women (82 per cent) returned to work part-time. Similarly to Neath Port Talbot, the majority of women (58 per cent) did not return to the same job as they held prior to the birth.

Figure 5.11   Age of child when mother returned to work

Age of child	West Dorset %
0-3 months	19
3-6 months	22
6-12 months	7
12-18 months	6
18 months-2 years	3
2-5 years	25
5-10 years	15
10-16 years	3

In comparison to Neath Port Talbot, significantly more women in West Dorset (58 per cent) said that having a child had affected the type of employment they undertook. This may be due to the more rural nature of West Dorset, which would restrict further the times a mother would be available to work as the school runs are longer and take more time.

Conclusion

In this chapter I have argued for the need to use both extensive and intensive research methodologies within research on labour markets and gender. Extensive quantitative data allows for statistical analysis on a representative sample, eliminating personal bias, something that is strongly favoured by policy advisors. Alongside this, intensive qualitative research methods allow for an in-depth understanding into attitudes, opinions and reasoning. With specific reference to my research on women's decision making regarding entering the formal labour market, extensive data will be drawn on to give an insight into Government practices and the implications of academic and NGO research on the formation of national policy. The understanding that national policy is based on aggregated national averages which rarely exist 'on the ground', raised the question of the extent of spatial variation at the local level. After discovering that spatial variation in women's labour market participation ranges from 40 per cent to 94 per cent, I have argued that it is necessary to use intensive qualitative techniques to further explore the reasons for this. In-depth, informal interviews with mothers in their homes, has enabled me to explore what factors are important to women when they are considering their participation in the formal labour market. I shall now go on to discuss the findings from these interviews. Although the following two chapters split the women up into those who work and those who do not work, the labour market history of each individual sometimes makes it difficult to draw such a clear cut distinction.

Chapter 6

Exploring the Interdependencies:
Women and Work

Introduction

The following two chapters draw on a series of semi-structured interviews undertaken with women in Neath Port Talbot and West Dorset about their labour market activity. For ease of analysis I have tried to simplify and categorise into two chapters what is in fact a complex network of interdependencies, which connect together to influence the geography of women's labour market activity. These two chapters are based around the twin themes of women and work, and women and home. Within each, I have identified the key influencing factors which contribute to and inform the decision to work or not work. Some of the issues underlying these factors were identified prior to undertaking the empirical research through an analysis of both academic reading on the development of feminist geography and labour market segmentation theory as discussed in chapter 2; and the current government priorities for women, as discussed in chapter 3, and so were specifically addressed in the interviews. This includes such aspects as childcare, work:life balance, the constraints of part-time employment and the need for flexibility from an employer. Other factors were identified later through the interviews themselves and appeared perhaps more important than previous academic analysis had allowed for, such as the influence of the partner's attitude, societal expectations and the decision that their mother had made regarding employment when they were a child. However, it gives a false picture to discuss the issues one by one in isolation. Throughout the reading of these chapters, what should be remembered is that every woman's view of labour market activity as a mother is multidimensional, and it is the varying combination of a number of factors which contributes to the way in which each woman views the formal labour market. Some of these interdependencies are generalised and applicable to the majority of women wherever they are, others are place specific. I shall identify the geographical differences as and when appropriate throughout the next two chapters, although they shall be discussed more explicitly in chapter 8 where there is an examination in more detail of how the complex interdependencies combine to produce different geographies.

This chapter concentrates on women's attitudes and opinions towards being a working mother. Seven main issues arose as key to influencing a woman's decision to enter the labour market once she has had children; predetermined expectations, the ease of returning to work, motivations for working, work:life balance, the need for flexibility, part-time employment and childcare. This chapter starts by looking at what predetermined expectations women feel once they have children regarding whether they should work or not work, and where that pressure comes from. Although it is initially only highlighted, the implications of the predetermined societal expectation on various aspects of women's lives will become evident throughout this chapter and the next. Secondly, I shall explore the ease with which women are able to (re)enter the formal labour market and the impact on and changes of the family routine which result. Thirdly, I attempt to summarise why women want to be working mothers, however, I acknowledge immediately that depending upon personal circumstances, every mother has a slightly different motivation. I recognise the influence of employment status in the decision-making process, with most women believing that women with 'careers' should not give it all up to stay at home as they will have worked hard to get where they are, and so employers should be more understanding of a working mother's needs. Fourthly, I shall look at the problems women face when trying to combine formal paid employment with their domestic responsibilities and discuss some of the demands working mothers have in order to achieve a work:life balance. I shall discuss the dilemma of guilt which all working mothers experience, asking themselves if they are doing the right thing by working whilst their children are still dependent. I shall go on to discuss one of these demands in more detail, that of flexibility from the employer, as it appears to be a key issue for the majority, if not all women, in order to be able to successfully contribute to the labour market at the same time as being a working mother. Sixthly, I shall discuss the possibility of part-time employment being the ideal solution for women, potentially enabling women to combine their dual role of mother and worker, giving them the flexibility to choose working hours to suit their domestic responsibilities. Finally, I shall look at the issues raised concerning childcare. The cost of childcare is often seen as the major barrier to women entering the labour market, but we shall see the lack of flexibility in childcare, as parents are expected more and more to work shift patterns outside of the normal 9-5 routine, is also critical.

Predetermined Expectations

As discussed in chapter 2, Tivers (1977) argued that the gender role constraint was *the* constraint that underlies all other influences on women's activity patterns (see also Currell 1974, Kirkpatric 1974, Richards 1988). Through my research I wanted to explore whether women felt there were any expectations on them to perform a particular role; that of full-time mother, or a working mum, how those expectations became evident, were enforced and what impact these have on the woman's decision to (re)enter the formal labour market as a mother. Having spoken to mothers in both Neath Port Talbot and West Dorset, all women felt that there were pre-determined societal expectations placed on them, with regard to career aspirations and working once they had children. However, expectations were not uniform for all mothers. Some felt they were expected to contribute to the family income by continuing to work at the same time as looking after the family. Others felt that there were low expectations of them in the formal labour market, either because it was assumed that career advancement would suffer as children became a priority and they would consequently not be interested in advancing their career (see Cockburn 1988) or it was assumed that they would give up work as soon as they had children to stay at home to be a full-time mum (see WGSG 1988, Bowlby et al. 1989 and Massey 1994). For example, D6[1] discussed how she 'was brought up to think she would always be a clerk typist or an office worker, nothing more was expected', and when she was pregnant with her first child,

> 'it was automatically assumed that I wouldn't work, by the people I worked with ... I felt that if I said I was going to go back, it would have been said that "surely you are going to stay at home and bring your child up yourself".'
>
> (D6, *working mother of two, had five years out of labour market*)

Although in almost all cases, the women had not had anything said to them directly regarding assumptions on what they should do, it is an unspoken matter which most women are highly conscious of, and feel the pressure from. Expectations and assumptions were not only felt from within the workplace from colleagues as described by D6, but as Walby (1997) and Holloway (1999) argue, discrimination based in gendered expectation also derives from many other sources such as: a partner's opinion; whether the neighbours work; the media portrayal of women including 'Hyperwoman' (Benn 1999, 104); and the overbearing traditional stereotypes of the woman as homemaker and the man as breadwinner. Moreover, through conducting my own interviews I also discovered that one of the predominant influences on a woman's decision was what 'choices' their mother had taken when they were a child. Although there were a few cases in both Neath Port Talbot and West Dorset where the woman had chosen to do the opposite of what they experienced as a child – for example D6 described how as a teenager she didn't like going home to an empty house, and so has tried to make sure that she is home waiting for her children when they come in from school – the majority of women in both areas enjoyed their childhood, and have in fact followed their mother's example.

> 'Her example of always being at home when we were young has made me the way I am and made me want to be at home with mine.'
>
> (D8, *mother of four, thirteen years out of labour market*)

---

[1] In order to maintain anonymity I have labelled the women interviewed with a letter (D=West Dorset, N=Neath Port Talbot) and a number (simply the order the interviews were conducted in each area). Any names mentioned in conversation have been changed.

'My mum was at home when I was younger and I thoroughly enjoyed it.'

(D13, *mother of two, works full-time from home*)

'The family had always worked, my mother worked, so I didn't think anything different.'

(N8, *mother of three, works full-time*)

'I got married in 1977 and had my eldest son in 1981. I gave up work. I had all intention of not going back to work again.'

(N14, mother *of three, twelve years out of labour market*)

The majority of women felt that they had 'turned out alright' and had seen the benefits from whichever decision their mother had made, and were of the opinion that, if their mother worked when they were a child, then they should too, or if their mother had been at home when they were a child, then they should be at home for their children too. Although when women were asked directly if they felt influenced by what their mother had done when they were a child, many appeared not to have consciously considered the link: such an influence did seem to be important, even if unconscious.

The partner was another important influence on the woman's decision to (re)enter the labour market. In all discussions with women about their employment activity, they mentioned the influence of their partner's opinion in their decision. For some mothers it was the insecurity of their partner's employment that kept them working, so that if their partner were to lose their job, there would still be regular money coming into the household to pay the bills. These were often women who had partners in the building trade or whose partners were self-employed, and who had previously experienced times of hardship from a reduction of income through a lack of partner's work. The income from the woman, in some cases only a small contribution, provided security that at least there would be some money coming in should the worst happen.

'I would love to have stayed at home, but the building trade is renowned for its periods of unemployment and we needed stability and certainty of wages coming in and my job provided us with that. From a family point of view it was going to provide the financial security, so that was the trade off. I would much rather have stayed at home, but it was never going to happen.'

(D18, *solicitor, mother of two, husband now at home full-time*)

More women commented on the influence of their partner's opinion than that of other societal expectations such as the media. Some partners were fully supportive of whatever the woman wanted to do, whether to stay at home with the children full-time or to remain employed in the formal labour market, which would involve negotiating childminding arrangements and taking on extra household responsibility. Others held stereotypical views – that they were the breadwinner and were supposed to support their family, and would feel as if they had failed should the mother have to go out to work rather than being at home to bring up the children. Although many women mocked these views as 'old fashioned', they adhered to them, and enjoyed the role they played within the family.

> 'He didn't want me to work, so I stayed at home...It was something in his mind, I should be home with the children.'
>> (N5, *mother of three, fourteen years out of labour market*)

One woman I spoke to in Neath Port Talbot gave the following comment on her questionnaire;

> 'My husband is a male chauvinist and believes a women's place is in the home and dinner should be on the table when he comes in.'
>> (N15, *mother of five, housewife*)

I was, therefore, keen to explore their relationship further and find out to what extent his opinion had shaped her labour market activity. N15 had worked as a secretary prior to marriage, remaining in employment until the birth of her first child, and re-entered the labour market after the birth of her second. After a nine year break to bring up her children, she went to college to bring her IT skills up to date, and returned to work within the same area as she had previously worked. The birth of her third child brought complications and so working was not an option as lots of time off was required; however, N15 remained active doing voluntary work wherever possible and then returned to full-time employment when her youngest child was in full-time education. The family then moved house, and although N15 initially remained in employment, she explained that the pressure of commuting took its toll and became too much and so she gave it up. However, irrespective of financial gains N15 clearly appreciates the social benefits of working and so she returned to voluntary work in the new area.

> 'I enjoy work, it is the socialising as well, it gets you out of the house for an hour or two and it gives you something different to focus on other than the children.'
>> (N15)

Moreover, as the conversation progressed, there appeared to be a lot more contributing to the strain of working than just the commuting, mainly her partner's 'traditional' views. Her partner retains the view that her priority is to be at home, and all domestic duties are her responsibility.

> 'The iron is an alien thing to him' (N15)

When the family first moved house, N15 explained that she applied for a job working in a nearby steel yard. However, her partner did not approve as he thought she would be surrounded by men all day, and he phoned up the company to inform them that she would not be attending her interview. Although N15 laughed about this incident, she displayed annoyance at this action, but described how she had 'come to accept it'. N15 admits that her partner has a demanding job which involves long hours, and he may not be home some nights till 10:30pm and, therefore, acknowledges that she is the primary carer for the children and must be at home to take care of them. However, all the children come home at different times and also want feeding at different times, all wanting different things for dinner. Additionally, her partner also expects her to have his dinner ready for when he comes home and so N15 finds herself cooking five different meals at five different times, even though she absolutely hates cooking.

> 'He doesn't like anything heated up, he likes it fresh, so if he comes in
> at 10pm, he wants it cooked then...you can't stick a curry in the oven,
> he likes a meat and two veg.'
>
> (N15)

Another aspect which N15 has to negotiate is the fact that her partner likes her to be in the house whenever he is there, and although it is okay for her to go out to get the shopping or pick up the children during the day, he does not really like her going out. N15 described how there was a film on at the cinema recently that she really wanted to see, but knew that he would not go with her and so she went without telling anyone during the day.

> 'Barry is not very keen on me going out...he gets very stressed if he is
> in the house and I am not here.'
>
> (N15)

Having spoken about the influence of her partner on not only her labour market activity, but also her daily routine, in rather derogatory terms, and continually displaying obvious signs that she was not happy with that situation. I went on to discuss with her how she now feels about being at home. Rather surprisingly, N15 enjoys her time during the day and believes that she is one of a lucky few who gets to do what she wants with her time and does not see her partner's attitude as a barrier to her activities!

> 'As long as everything is done and as long as I am here when Barry
> comes in, the day is mine, I can do what I want....I don't think there are
> a lot of people who can do that and have the freedom I have got...I think
> I am lucky compared with a lot of people.'
>
> (N15)

Although I believe N15 in many aspects to be an extreme case, the partner appeared to be a fairly strong influence for almost all the women I spoke to. Their partners' opinion and ability to support the women in their decision would influence whether or not a woman was able to (re)enter the labour market, and subsequently the type of employment the woman was able to undertake, through being willing to share the domestic responsibilities. As the WGSG (1984) argued, if the partner is prepared to share responsibility for domestic chores and childcare, then the mother would be able to (re)enter into the employment of her choice. However, if the partner is either unable or unwilling to take on equal responsibility for bringing up the family and looking after the home, then the mother is restricted to taking employment which fits around her existing routine. Reasons for the latter ranged from the partner having a particular type of job, such as D19 whose partner was a retained fire-fighter and was on call 24 hours a day 365 days a year (meaning that he could never be with the children on his own in case his pager went off), to maintaining 'traditional' views that it was the mother's responsibility to bring up the children, so if she wanted to enter into the labour market, it was her responsibility to find a job which fitted with her domestic responsibilities or to organise and pay for the childminding.

I spoke to a variety of working and non-working women in both Neath Port Talbot and West Dorset, and there was a general division between the two areas in societal expectations. Whilst I have described the conversation with N15 as extreme, it is by no means unusual in a locality such as Neath Port Talbot. Repeatedly within the Neath Port Talbot area, women expressed their belief that they are expected to give up work and stay at home once they have children.

> 'I think because I automatically assumed I would get pregnant and never work again, I gave up a very good career in the civil service.'
> (N14, *mother of three, had twelve years out of labour market*)

Many women felt as if the assumption that they should stay at home had arrived from past generations, echoing Massey's (1994) thoughts on the development of the male breadwinner, female homemaker culture. Neath Port Talbot was developed and dominated by the coalmining industry where social culture was centred around men with working men's clubs and male voice choirs. As Humphrys (1972) and Massey (1984) discovered in their explorations of South Wales, women in these areas were often in their home due not only to the lack of employment for women in the single industry area, but also to the nature of the male employment which demanded large quantities of domestic labour around the clock as the women looked after not only their husbands, but also their sons. Since these often worked different shifts, all the men of the house would want feeding and washing at different times. Even though the Neath Port Talbot area has undergone major social and economic changes since the closure of the coalmines, the stereotype of male breadwinner, female homemaker is still evident, not only amongst the older generations who were part of the mining period, but also some of the younger generation, as described by N9,

> 'I think it is very hard for men to see you in a different role, other than the one at home. I think my husband tried very hard to be a man of the 90s, but he still says things like "I've washed the dishes". He thinks he has done it for me, and I am constantly battling to say "it isn't my role anymore". He did something for me yesterday, he went to my sister's to

borrow the sewing machine. He said to me "you didn't even say thank
you", and I felt like saying have you said thank you for all the washing
and ironing I have done all weekend. I still do that, I still take
responsibility for those things. Sometimes he is home before me and
does tea, but he thinks then he has done you a favour, rather than taking
responsibility. I find that quite hard. I have spoken to quite a few
women who are working and they come up against the same thing and I
think it is a hard thing to overcome, they just can not see it is not your
role anymore.'

(N9, *mother of two, returned to work full-time*
after eleven years out of the labour market)

In conversations with women regarding the decisions they had made, many
women compared themselves with other women they knew, for example, friends
and neighbours, as a way of justifying what they do. Two women in Neath Port
Talbot who were in employment explained how they knew they were not following
the norm, and were looked down upon by other women for not doing so.

'Down at the school, I am looked down upon because I work. It is 'that'
lot from the estate down there, I am working to pay their benefits, so I
don't know why they look down their noses. Some of them are lucky,
their husbands are much better paid and they don't have to work, but
some of them know how to work the system. They are claiming
benefits, but they are working, so they are bringing in more than me, but
it is all under the counter.'

(N8, *mother of three, returned to work immediately after each child*)

'Most of my friends don't work, I am about the only one out of a group
of about 15-20 women.'

(N16, *mother of two, eleven years out of labour market*)

In contrast, although Neath Port Talbot is a poorer area than West Dorset, the
majority of women in West Dorset talked of the expectation that they would
continue to work once they had children. Although again pressure came from a
combination of other mothers, partners and society in general, some women also
mentioned financial pressure, not because it was necessary to work in order to
provide the basics as in Neath Port Talbot, but the women in West Dorset wanted
money to provide the extras, the luxuries which enable them to conform with
others around them.

'It just feels like pressure from everybody, I know my neighbours have
tittle-tattled about my neighbour who doesn't work and she has only one
kid, so I thought I don't want people thinking that and seeing my
husband working all the hours and getting really tired, I need to pay my
way and contribute.'

(D7, *mother of four, ten years out of labour market*)

'Your position you get yourself into is your choice. I think people are pressured to a certain type of life which they can ill afford sometimes, through the media and peer pressure and parental pressure to conform.'

(D12, *self-employed mother of one*)

Some women indicated that the Government pressurise women to work through the promotion of initiatives to get women into work such as Working Families Tax Credit, yet they also publish reports which are brought to the publics attention through the media, that tell you your children suffer if you work whilst they are young as demonstrated with a few examples in figure 6.1.

Figure 6.1   Newspapers present women with mixed messages

A working mum is just the job, say children
Lisa Buckingham, Saturday 11th October 1997, *The Guardian*

Working mums guilty (again)
Catherine Bennett, Thursday 11th November 1999, *The Guardian*

Children of working mothers 'at risk'
John Carvel, Wednesday 14th March 2001, *The Guardian*

Childcare study boosts back-to-work mothers
Maureen Freely, Sunday 30th April 2000, *The Observer*

Child study finds working mums do no lasting harm
Julian Borger, Tuesday 2nd March 1999, *The Guardian*

Children of working mothers lag behind
Sarah Harris, 30th July 2002, *Daily Mail*

Mothers working full-time 'damage families'
Parenting, 13th June 2001, *Daily Mail*

'I am choosing not to want it all'
Elizabeth Grice, 3rd February 2003, *The Daily Telegraph*

'We love our children too: Despite what the critics may suggest, working mothers are not child abusers'
Maureen Freely, 6th February 2000, *The Observer*

'Phew, working mothers can relax – until the blindingly obvious strikes again'
Annie Ashworth, 14th August 2003, *The Times*

As a result of these mixed messages, women tend to ignore the Government and any campaigns it has to encourage women into the workplace.

> 'This whole New Deal for Lone Parents has got a big part to play. The way they have geared it, it does make women feel guilty if they are not out there working. I think there are mixed messages coming through as well.'
>
> (D9, *mother of three, unemployed, actively seeking work*)

> 'I think it is very confused. The media is trying to say that they should be at home, but then in the next breathe they are saying they should be able to do what they want and have it all and ignore the practicalities. The misleading thing is that it looks like you can have it all and nothing will suffer, well that is obviously rubbish.'
>
> (D15, *mother of two self-employed accountant*)

> 'I would really like to see some research to see if it is better or worse for the children. Nothing that you read substantiates one of the other. It all contradicts each other. Something comes out which says you should stay at home and something else says they are better if you work.'
>
> (D16, *mother of three, midwife*)

A couple of woman I spoke to in West Dorset believed that there is incredible pressure on women to work from society in general. They both talked in detail of the number of times they had been at dinner parties with friends or partners work colleagues, and they feared the moment of having to tell people that they were at home full-time with their children for they knew the reaction would be a negative one.

> 'It is always that "I am only a mother" phase which I hated it is a conversation stopper.'
>
> (D15, *mother of two, had three years out of labour market*)

> 'When people ask me what I do, I say I stay at home to look after the kids and it is a real conversation stopper. They think you are a nobody, you really are a nobody, even though I feel I am doing a proper job, people only judge you by what you do.'
>
> (D5 *mother of four, at home full-time*)

Rather than the woman at home being the norm as in Neath Port Talbot, women at home in West Dorset feel that they have to justify their decision and make many sacrifices in order to be able to do so. They constantly make comparisons with other people they know, and are conscious that they do not go on holiday abroad, they only have one car, and they cannot afford to provide the extravagant birthday parties for their children that their friends do. However, for them the benefits of being at home full-time with their children and watching them develop far outweighs the materialistic gains of working.

'You get pressure from your husband because he has friends who work and they get holidays and things like that. We don't go on holiday.'

(D5, *mother of four, at home full-time*)

I shall discuss the issues surrounding women at home in the next chapter. In this section I have highlighted the differentiation in expectations which women face in the two case study areas. Although there were women in Neath Port Talbot who felt they should be out at work, and some women I spoke to in West Dorset felt that they should be at home with their children, the majority of women in Neath Port Talbot described the societal pressures they felt indicating that they should give up work to be at home full-time once they had children. In contrast the majority of women of West Dorset indicated that they felt they were expected to remain in work at the same time as looking after the family. I have only touched the surface of the expectations placed on women, and am aware that I have not discussed the repercussions of such expectations. However, these will become evident throughout this chapter as I look at other factors which affect women's decision to enter the formal labour market. This will be discussed in further detail in chapter 8.

The Ease of Returning to Employment

Returning to work after having a child can often be a daunting event, whether straight after maternity leave when the mother worries about leaving a young child or after a long break from the labour market when she has been at home with her children for a few years, and she may no longer be there when they come in from school. Many of the same concerns arise whether the woman is returning to a familiar job or new employment, full-time or part-time. For all mothers, the primary concern is whether the child(ren) are going to be okay, if their routine is going to change. Once the concerns over the child(ren) are managed, then the woman may worry about her ability to undertake formal employment if her skills are out of date, or whether she will be able to retain her status and still do her job if she only works part-time.

Several of the women in both Neath Port Talbot and West Dorset had taken time out of the formal labour market in order to be at home with their children during the early years, but had subsequently re-entered employment. I was keen to explore the ease of transition from full-time mum to working mum, not only in terms of practical changes to the daily routine, but also with regard to the change in character as the woman assumes a role outside of the family.

More women in Neath Port Talbot discussed the problem of finding jobs in the area, more specifically suitable employment to fit in with their requirements. By 'suitable' the women are referring to employment which they would be able to do without adverse impact on their ability to take and collect their children from school. For this reason many women were not able to work in factories as this often meant shift work over unsociable hours. Due to the high number of low-skilled employees in the area and the low number of jobs, even those women who would like to consider entering the labour market are not always able to do so.

> 'There are no jobs in this area, and by the time you have paid petrol, well I might as well stay home for 30 quid or whatever...Like I went to work for Asda in Aberdare when I was in college, only part-time, only £80 a week, not much. It was crazy there were about 4-500 people getting interviewed for 30 jobs, it was crap here.'
>
> (N16, *mother of two, has returned to work after*
> *eleven years out of labour market*)

Irrespective of how long the woman had been out of the labour market, on returning to work, the women did not feel that there had been too many changes in their daily routine. There were only two factors which were mentioned that had to change; firstly, there had to be a more organised routine in the morning in order to make sure that everyone was ready to leave together; and secondly the partner had to take on more responsibility within the household; for example, help more with housework, or cook for the children if they were home first.

> 'All simple things, the morning had to become a strict routine, instead of everyone getting ready and me taking the kids to school, I had to leave a lot earlier. They are usually the first ones in school. Before it didn't matter if I wasn't ready, I could just have a bath when I came home, but now there is more of us to get ready. Both Rachel and Alison have had to cook meals for me sometimes if I have gone straight to college from work, or been late because of training courses. They have to help out a lot more. When I wasn't working you would have the whole holidays to have pure fun, we did last holiday, but now I ended up catching up with all the housework. Here are my kids at home and I was doing housework.'
>
> (D7, *mother of four, had ten years out of labour market,*
> *returned to 30 hours a week*)

> 'I think the person who saw the difference the most was my husband because he had to take much more of a responsibility, and he found it quite a struggle...It was a constant battle...it was a hard battle to get him to see that I was tired and I needed his support and needed him to take a share in the chores.'
>
> (N9, *mother of two, ten years out of formal labour market, went to*
> *college, now in profession*)

The reason for the minimal change in routine was often because either the woman had only accepted a job which would cause minimal disruption to an already established routine, or the woman had previously been in education and the routine of mum needing time to herself had again already been established and taking on employment was just the next step. Most women felt that as a consequence of the mother (re)entering the formal labour market, and the partner having to become more involved in the daily routines of running a household, there had been a positive effect on the family. Many women described how closer bonds have grown between the father and the child(ren), and how the child(ren) became more sociable as they have gone to a childminder and have been forced into mixing with other children.

> 'I think if anything it has made a stronger bond between the children and him. Up till then, because I was always there, I did it. I was the one putting them to bed and things like that. He would be here, but they have a far stronger relationship because they have spent more time with him.'
>
> (N14, *after retraining and taking a professional career*)

> 'The positive side to me working, is because he [her son] has had to go elsewhere, he has had to become more confident. That is a really big bonus because he was a shy little lad. Over time through going to school, he has slowly gained confidence because he is mixing with others and has had to be picked up by different people.'
>
> (D7, *working mother of four, returned after ten years out of labour market*)

When initially considering a return to the workplace, some women felt unconfident about their level of ability, and their lack of skills, particularly IT and computer skills, and many decided to go back to college to retrain or update their skills. Other women, who did not have any initial worries about returning to work, only realised how unconfident they had become, once they were back in the working environment.

> I didn't realised how unconfident I had become until I got out and realised, oh god I am a bit of a quivering wreck.'
>
> (D8, *mother of four, had 12 ½ years out of labour market*)

It may have been expected that women (re)entering the labour market after time out would require dramatic changes to the household. However the women I spoke to in Neath Port Talbot and West Dorset, managed to return to work without too many problems. This is often because they chose jobs which they could fit around their domestic requirements or they eased themselves back in by first taking on part-time work to see how they managed whilst the children were still young. Many women then returned full-time once the children were older, and they felt more confident that they were able to dedicate themselves to the demands of the job. I shall discuss issues surrounding part-time employment later in this chapter.

Motivations for Working

All the women I spoke to in both Neath Port Talbot and West Dorset who were in the formal labour market, talked of the many ways in which they, and their family, have benefited from them being in work. Some women spoke of the obvious financial benefit of them being employed in the formal labour market; for some, it was the ability to have an independent income rather than being entirely dependent on their partner, for others, it was the ability to be able to provide the extras such as family holidays, school trips and 'the latest trainers'. However, money was by no means the predominant reason for entering the labour market, often the money was an added bonus to other social benefits.

> 'Although it helps a bit financially, my wage is not that good; the main benefit for me is that I am getting out of the house. Some weeks I don't earn £30, and perhaps I have been out 40 hours, but then at least I am not stuck in the house, I enjoy what I am doing.'
>
> (N5, *delivers catalogues*)

As N5 describes, women find there are many more important beneficial factors to working other than money; there are social gains such as having something else to focus on other than the children, getting adult stimulation, having a sense of purpose outside of the home, being valued, making new friends, gaining self-worth, and gaining confidence. These social benefits were particularly noted by women who had had a few years out of the labour market in order to be at home full-time to bring up their children. For many women, due to the fact that they were used to living on one income, money was not the main reason for their return, but rather the social benefits the women gained by having an external focus outside of the family home. Although the mothers had a wonderful time with their children and would not have changed their decision to be at home, when the children were old enough to be in school full-time, the mothers were often longing to go back to work. As Mackenzie and Rose argue, many women take jobs to escape the boredom and isolation of the 'ideal housewife' role (1983, 179). Whether the woman worked only a few hours a week, or went straight back into full-time employment, all the women discussed how not only had they benefited, but also their relationship with their partner had improved as they had something other than the children or housework to talk to them about. Moreover, this had a knock-on effect on their confidence and level of self-worth.

> 'I think that when you are not working you become so wrapped up in your children, they are such a part of your world. In can be unhealthy as your children are so important and you lose the sense of reality that they are only kids, and you think your kids are the best in the world, but they are just kids and you need to get away from that to realise, you have to separate yourself from that.'
>
> (D6, *mother of two, five years out of labour market*)

'I haven't gone mad and killed the children off!! I wouldn't have gone that far, but I think for my personal well being I think it has made a huge difference. I think that the children also got more socialisation because I was working.'

(D13, *mother of two, works from home*)

Many women referred to the status of a job as an influential factor on their decision making regarding whether or not to give up work once they had a child. This opinion was expressed both by women in professional jobs who described that they were reluctant to give these up as they had trained hard for a long time to get there, and by women who had given up, who expressed that had they been in a professional career, they would think twice before giving it up. Both groups of women referred to shop work and factory work as menial and something which could be given up easily.

'I could have given up, but because I had trained for four years and I was a professional...it would have been silly for me to give up...I had a career to go to. If I was just working in a shop perhaps I would have given up.'

(D6, *teacher*)

'I do think it is important to be home with them, unless you are in a career job. You can understand people like that who are high up, reluctant to give up their work, but when you are doing menial work.'

(N3, *mother of two, part-time worker as night sitter in residential school*)

'If I was just working in a shop perhaps I would have given up.'

(N13, *working mother of two, teacher*)

Interestingly, the status of the job was referred to as an important decision influencing factor, more often than other benefits of working such as independent income, intellectual stimulation or simply 'getting out of the house'. What should be noted is that all women who felt they were not in a job where they were able to drop everything should a situation arise to do so, had flexible childcare, family and/or friends nearby that they were able to call on in times of need.

Many women who had had a few years out of the labour market to be at home with their children and who were (re)entering the work place, were not necessarily interested in a 'career' job, but simply undertook employment for personal social benefits such as adult interaction, 'getting out of the house', meeting new friends, being appreciated and having an independent income.

'Meeting other people and friendships... I met different people who were nothing to do with the church. Nothing to do with the family and nothing to do with Steve's work, so that was really good... I wanted a change, that was definitely the first reason for it, and the break from the children.'

(D8, *mother of four, had 12 ½ years out of formal labour market*)

> 'It is the gratification, being appreciated. Your family don't appreciate
> you, it is an ego thing. If you go in and someone say 'I couldn't have
> done it without you', and you say 'oh no, you did it all', but really there
> is that part of you going 'oh yes!'. We all need a pat on the head and if
> you can't get it at home, you go and get it elsewhere. I get mine from
> work.'
>
> (N8, *working mother of three, midwife*)

In both Neath Port Talbot and West Dorset, it is true to say that those women who had trained and entered a professional career were keen to maintain their employment status after the birth of their child(ren), more so than women who were not in professional employment however, in the majority of cases this could only be achieved if the women remain part of a stable family unit. If the family unit broke up, then the woman was no longer able to maintain her professional employment, either due to the lack of support available to share the domestic responsibilities, or no longer being able to afford the childcare to enable her to do so. This indicates that although employment status can influence the woman's intentions, it may not always be possible to follow them through.

Not all the benefits of working were directly related to the women themselves. Besides the financial gain for the family and the social gains for the mother, some women also felt that they were providing a good role model for their children by breaking the stereotype of 'woman in the home'. As previously described, it became evident to me that the general feeling for women in the Neath Port Talbot area was that they would give up employment and stay at home to be full-time mothers. Those mothers who were in work, felt as if they were breaking the norm and giving their children a broader outlook.

> 'I think it gives them a different outlook as they both talk quite freely
> now about when they go to university, and I think that the valley culture
> is that you don't go to university, but they are talking matter of fact
> about it. I think it has been a positive aspect, and they see the woman's
> roles now as not just being housewife. I don't think I had that
> confidence or outlook when I was their age.'
>
> (N9, *mother of two, 10 years out of formal labour market, went to
> college, now in profession*)

In summary, although some women mentioned the enjoyment of being able to provide security or 'extras' from their employment in the formal labour market, the majority of women worked for the social benefits of having an external focus, meeting new people, intellectual stimulation or simply 'getting out of the house'. I shall now go on to look at the demands on women who try to work and fulfil their domestic role, and whether they are able to establish a work:life balance.

Work:life Balance

With the introduction of the new knowledge-based economy, the pressure for women to join and remain in the labour market has dramatically increased in recent years. As a result, enabling women to maintain a work:life balance is a key policy challenge of the twenty-first century, for trying to establish a work:life balance is an issue which causes many women anguish, because they feel that in order to try and do one task to the best of their ability, they have to neglect the others. It is often a constant battle of swings and roundabouts, for at different times, different events take priority, albeit the children, work, the home, education or simply time alone. This constant battle can often bring with it feelings of guilt that they were neglecting one of their responsibilities; either they felt they were letting their colleagues down if they had to take too much time off; or their family if they were not there to keep the family home clean and tidy; or the children if they were not there when they got home from school to help them with homework; or to hear them read. This endorses the arguments of Hakim (1993, 1995, 1996c) and Benn (1999) who believe that there can never be a true work:life balance. Benn describes how women try and be part of two worlds, that of mother and career worker, but in fact are never fully present in either, and Hakim argues that you are either a 'careerist', or 'home-centred', you can never be both.

Even if mothers have many good reasons for entering the labour market, such as wanting the extra income or gaining the social benefits of working, they question whether or not they are doing the right thing for their child(ren). Mothers worry that they are neglecting their child(ren) by not taking them to school, picking them up and being home with them when they come home from school. Working mothers mentioned aspects such as helping with homework, keeping the house tidy or having time to hear the children read every day as activities which routinely get neglected.

> 'I want to listen to my kids read. I looked through the reading diary, that was another thing I found, when I was at home I was listening to the little ones every single day and I thought I was still doing, but I looked in the reading diary, and no. Sometimes I was just too tired, too busy, or I have to go to college.'
>
> (D7, *working mother of four*)

> 'Sometimes I feel guilty that I am working. I think I am their mother and I should be doing for them, but I get mixed feelings. I think they are older now and they should be doing for themselves, but sometimes they want certain clothes to go out to wear and I haven't washed it. Very much guilt. That is a very big part of it.'
>
> (N7, *working mother of two*)

The feeling of guilt is particularly prevalent with pre-school children, the women worried that they were not giving enough time to their child(ren) to help them develop, and were upset that they were missing out on them growing up. Moments such as the first steps, the first words and the first day at school, were all poignant moments in the child's life that made the woman rethink her working situation.

> 'I didn't get to see Rhiannon growing up, guaranteed she would do something new when I wasn't there.'
>
> (N18, *working mother of two*)

The way in which mothers appear to deal with their guilt is either by giving up one of their responsibilities for example, no longer being house proud or employing a cleaner so that you have more time with the children. Many women felt the need to prioritise, which often meant the children came first at the sacrifice of her own desires and in a couple of cases, their relationship with her partner. In order to provide the children with the upbringing the parents wanted, sometimes, the mother and father would no longer have time with each other. For example, N18 described how she and her husband were both in the nursing profession and had arranged to work opposite shifts in order that there was always someone home with the children. This meant that she and her husband were hardly ever at home together apart from a week a year where they overlap their annual holiday to be together as a family.

> 'My husband and I hardly ever have time off together, he was just somebody else who was in the house to share the childcare. We found it easier to work opposite each other to be around for the kids.'
>
> (N18, *working mother of two*)

Alternatively women began to accept that they cannot do everything all of the time, but they were doing the best they could.

> 'Even though I don't think I am doing anything well, I am doing everything alright.'
>
> (D3, *working mother of four*)

Although the issue of guilt came up with almost all women I spoke to in both Neath Port Talbot and West Dorset, it was also something which mothers tried to forget or bury under the pile of positive effects. The women discussed how they had thought about the alternative of staying at home, but decided that they needed or wanted to work, and they would just have to deal with the consequences.

Those women who felt they had managed to maintain a work:life balance whilst being a working mother declared the secret to success as determination and organisation. These women believed that if you wanted to make it happen, if it was important for you to remain in employment, then you make it work, 'you *have* to make it work, no one else is going to do it for you' (D1). Mothers explained that forward thinking and preparation mean that time can be used efficiently to ensure everything can be fitted in.

'Even though he hasn't got swimming till Thursday [*interviewed on a Sunday*] I have already got his trunks and things ready. It is thinking in advance.'

(D3, *working mother of four*)

'We are good at using time early in the morning or late at night... But it did get a lot when I had Robert young. If I was away [*on business*] and he wasn't coming, he used to go to granny's Sunday night, taxi from granny's to school. He would go home to the assistant in the shop and then stay at a friends on Monday night. I would organise all this before I went away. That was an art in management. It functioned as long as your forward planning was good and nothing happened.'

(D12, *self-employed antique dealer*)

D12 explained to me how for most of her life she has been a self-employed single parent, which has had just as many perks as problems. Although it has been tough as she has had to do everything on her own and the business was not always done well, being self-employed also meant that she was able to control her flexibility and organise her employment needs with her domestic responsibilities. D12 runs a shop through which she buys and sells antique furniture, and when she needed to take her son to and from nursery or school, she simply had to shut up shop for a while.

'I used to just leave everything and race back...I had no help.'   (D12)

Although most women appreciated the benefits of working as well as bringing up children, there are also the negative aspects of being a working mother. Some mothers expressed how the stress caused from trying to create a work:life balance and ensure that the children were given everything they needed (time and help, not materialistic) has caused the women to be exhausted. The majority of women in both Neath Port Talbot and West Dorset felt that they were the primary carer for their child(ren) and it always came down to them to ensure that the children were in the right place at the right time and had everything they needed with them, including such activities as washing the PE kit the night before. Those women who had never considered staying at home were not resentful of those women who were not in employment, they just acknowledged that they simply had to get on with it, and that was life.

Flexibility

One of the biggest issues, which women consistently raised as a pressing concern for them when making their decision on employment, was the need for flexibility. Women felt that their family would always have to come first and employers should be aware of that and perhaps be more understanding of the difficulties when problems arise. For it does not matter how organised a mother is, and how much planning she is able to do in advance, there will always be the occasional day when things do not go to plan, and the mother then needs flexibility in her employment to be able to deal with the situation.

> 'I wish that employers would consider, when they take on a woman of a certain age, just look at it as a whole picture rather than just a female, then give priority on things like holiday dates, within reason, so that the summer holidays are not so much of a nightmare.'
> *(working friend of D4, single mother of two)*

For some women, this may involve a renegotiation of starting hours so that they are able to take children to school on the way to work: for other women, they may want to work flexi-hours, so that they can take time off for such times as doctors'/dentists' appointments, without it coming from their annual leave. Many women talked of wanting to work term time only so that they can be home with their children during the holidays, or a compromise to this may be to negotiate annualised hours so that they are able to work longer hours at other times of the year, and take time out during the holidays without it affecting their annual leave allowance. D6 discusses how her employers have been flexible to allow her to work when it suits her.

> 'I have increased my hours and discussed it with them and they agreed that I would never have to work during school holiday times. I said my children had to come first...They did not mind as long as I worked the hours for the year, it didn't matter how I worked them...It gave me spare days, so if there was a problem with my husband, I could juggle the days by having a different day in the week off. They are quite happy with that as long as I work the number of days I am supposed to.'
> (D6, *mother of two, returned to work after five years, working term time only, later negotiated annualised hours*)

However, in contrast, a friend of D4 describes how her employers have noted her priorities and have reduced her flexibility by formalising her working hours.

> 'I have just had to renegotiate my contract. I used to be able to go into work after dropping my daughter at school, but as time has gone on they have realised that is 15 minutes a day, every morning. They have now rewritten my contract so that I start at 9:30 so I have in effect lost 2.5 hours pay a week because of that, but I am peed off about it because I never take a lunch, so to me it is swings and roundabouts, but they didn't accept that. So now, I take a lunch, I make a point of it. They have been so inflexible and cut my pay which is the last thing I needed.'
> *(working friend of D4, single mother of two)*

Although there are some occasions when a temporary alteration in hours can be negotiated in advance, for example school holidays, sometimes eventualities occur which cannot be planned for, such as when the child is sick or the organised childcare lets them down at the last minute. Each of these occasions places added pressure on mothers to be flexible, sometimes with little notice. Each working woman I spoke to was able to recount for me several occasions when they had needed a bit of flexibility from their employer on such occasions but had been denied it, causing great difficulties.   Inset[2] days cause a lot of problems for mothers, as schools often only inform parents a couple of weeks in advance, making it difficult to make arrangements. Mothers with more than one child, explained how at some point the children are at different schools, and so they have different inset days, causing more difficulties.

> 'Those are murder. Inset days…very often they give me only a couple of weeks notice and it is too late to swop shifts and it can be a problem.'
>
> (N18, *mother of two, senior nurse*)

> 'I have actually written to my employers this week giving them 4 weeks notice saying that I am really stuck, but they won't let me have the day off.'
>
> (D4, *working mother of two talking of an inset day*)

As mentioned above, an additional dilemma women face with their employment has to be addressed when their child is sick and mum needs to be at home. Do they phone work and explain the situation hoping that their employer will be sympathetic or call in sick themselves? For many women, the answer to this question has come through trial and error.

> 'If the children are sick, you are stuck because you are not supposed to take the day as holiday and that is the biggest bug bearer, when you think you have it all sorted out and you can't do it because the children are sick… It is not you who is sick, it is your job and if you are sick that is okay, but not your children.'
>
> (D16, *working mother of three*)

> I had had one day off to look after my daughter when she was unwell when she was a year old. I had one day off and the colleague that I work very closely with told me that another colleague had made a comment that I shouldn't be having time off to look after my children. That I shouldn't be working and my job was to be at home… . The interesting thing was …, the female colleague who told me about it said "Next time you need time off for Erica, just say that you are ill". That would have been alright, take the time, don't tell anyone it's for your daughter and then everyone is happy.'
>
> (D2, *mother of two, teacher*)

None of the women were asking for additional time off or special favours, simply a bit of understanding and a degree of flexibility from their employer to allow them to meet their family demands and be able to catch up the missed hours another time. Those women who felt that their employer had been good to them, allowing

---

[2] Teacher training days where children are not allowed into school. There are approximately five days sporadically organised throughout the school year.

them the flexibility they needed when they needed it, in order to maintain their employment at the same time as carrying out their domestic duties, all expressed strong loyalty to their employer. The women were very appreciative of the flexibility and recognise that they are 'lucky' compared with other women. As a consequence they became a loyal employee and are happy to stay where they were, even if that meant passing up a promotion opportunity elsewhere, for their employer.

Although many women feel that employers should be sympathetic towards mothers and allow them the flexibility to work their jobs around their family commitments, they are also aware that sometimes this can cause resentment from colleagues who do not have children. In a conversation with one mother who was a nurse, she had thought about the possibility of working hard and doing all her hours in two or three days so that she could reduce her childcare costs and have four days at home with her child. Although this was a realistic possibility as other members of staff were working a similar arrangement, she was aware of the resentment of her colleagues (those who were not mums or whose children were old enough to be in school) who were restricted to working particular shift hours of 7am to 3pm, or 3pm to 9pm. Although it is necessary to have core hours of cover, the presence of resentment amongst colleagues highlights the need for flexible arrangements to be made available to all employees, men and women, mothers and non-mothers, in order that people feel they can utilise the flexibility rather than not utilising it from fear of resentment from colleagues.

Many of the women in both Neath Port Talbot and West Dorset noted that often a company will promote themselves as 'family-friendly' and in theory, accept the idea of flexible working. However, many of the mothers I spoke to said that it is often the attitude of their boss which determined whether or not flexibility is applied, for if the boss is not understanding of the juggling act which is required with children, they are unwilling to be flexible with working arrangements.

> 'My boss at that time, I don't think it is too strong to describe him as a misogynist, and he disapproved of children. He was quite knurled, when I told him I was pregnant, he told me that I was out of my brain, I was not going to be much use to him after that. He could not talk to me for months. Even after I had had Harriet, it was a small office, only 16 people, so everyone got to know everyone's partners and children's names, but he always called her 'the baby' he could never call her by her name. So when I said 'can I go part-time', it was going to have work implications for him, recruiting was a problem and in his previous life, job shares had not worked, so he was against that. So basically he said no.'
>
> (D18,*working mother of two, solicitor*)

> 'They talk about being a family-friendly environment, but that is a load of codswallop really.'
>
> (D1, *working mother of two, nurse*)

Some women spoke of being in a job where flexibility is restricted by the nature of their employment. A few women in both Neath Port Talbot and West Dorset were in nursing careers or tourism-related employment, and the demands of their jobs restricted the level of flexibility they had due to commitments at work.

> 'The job that I am in is holiday tourism-based and I am not allowed to take holidays during school holiday time, which is a nightmare. I am allowed one week in the school holidays which is the last week of the holidays which is the quietest time. I only agreed to taking the job if they gave me that week, and they reluctantly did.'
>
> (D4, *single mother of two*)

Although they could perhaps negotiate regular hours to suit their domestic duties, they would not be in a position to 'drop everything', should a last minute situation arise. I shall discuss the ways in which the level of responsibility and employment status affect women's ability to maintain their domestic responsibilities later in this chapter.

Some of the women I spoke to were aware that the government is trying to help them remain in work through the promotion of 'family-friendly' policies and such initiatives as Working Family Tax Credit and Children's Tax Credit. However, they argue that their policies do not necessarily ensure that flexibility actually happens in their work place, particularly if they work for a small company where the pressures of paying WFTC contributions are greater. Most women believe that the responsibility for flexible working lies with the employer. As D4 expresses,

> 'I think, the government are doing things and that is all well and good, but I think the pressure needs to be put on the employers to have a little more of an understanding. The government can put in the money and the places, and yes we do need that, but if you haven't got a sympathetic employer, it is not going to work anyway.'
>
> (D4, *working mother of two*)

Flexibility within employment is key to whether or not women are able to enter into the formal labour market at the same time as maintaining their domestic responsibilities. This is of equal importance for women in both Neath Port Talbot and West Dorset. In order to increase their flexibility many women (re)enter the labour market in part-time employment believing this will give them the 'best of both worlds'. I shall go on to discuss whether in fact that was true for the women of Neath Port Talbot and West Dorset.

Part-time Employment

As Walby (1997) argues, women choose to work part-time as a way of balancing existing commitments to employment and to home life and it can therefore be seen as the ideal way to combine the social benefits of working along with the opportunity to be at home when your children are there (see also Beechey and Perkins 1995). Part-time employment can range from just a few hours a week to a few hours a day, with most mothers seeking employment during school hours so that they are still able to take the children to school, and pick them up at the end of the day. Many of the women who were employed full-time commented that the one thing they wished they were able to do, is to be waiting at the school gate to pick up their children with all the other mums. Whilst a child is young, part-time employment can be used as a stepping stone from being at home full-time (either during maternity leave or having had a break from the labour market) before returning to full-time employment.

> 'I didn't want something that was going to be too demanding because obviously my family came first. I always thought they came first and that is why I went part-time.'
> (D6, *mother of two, had five years out of labour market*)

In theory part-time employment provides the 'best of both worlds' however, for some women part-time employment can cause its own dilemmas as they struggle to give 100 per cent to both work and their family. As discussed in chapter 2, Hanson and Pratt (1995) believe that women working part-time move within particularly constrained spatial orbits and they struggle to balance their dual roles, and they are never fully part of one world or the other. This was a concern expressed by some of the women I spoke to, who originally thought that working part-time would be a great way for them to combine the need to work with the ability to look after the children, but later discovered that it was not that easy.

> 'I thought it would be really easy, working 30 hours I thought it would be really simple, but it has been unbelievably stressful...I have enjoyed going back to work, but because of the hassle with the kids, it has ended up being more negatives than positives.'
> (D7, *mother of four, had ten years out of the formal labour market*)

> 'If you are working part-time and you are paying for clubs after school, it is not worth it.'
> (N3, *mother of two, had twelve years out of formal labour market*)

Those women who remained working in order to maintain their career, noted that working part-time can also limit the opportunities for career advancement as assumptions are made that if the woman is working part-time due to family commitments, she will not be interested in taking on more responsibility or promotion. Women who worked part-time felt they were overlooked for promotion opportunities and were not offered training courses. Additionally, mothers working part-time may also be seen by their employer as unreliable, for mothers often remain the primary carer for the children and are therefore the first emergency contact if there is a problem with the family and will require time off work at short notice; something which is not seen as conducive to professional employment. Although the Government introduced the Part-time Directive in 2000 to ensure that part-time workers were given the same rights and access to opportunities as full-time employees (discussed in chapter 3), the women I spoke to found that if they reduced their hours once they had children, they lost status at work, either in terms of respect from work colleagues, or in their expected career aspirations.

> 'People assume that when you are pregnant and you come back from maternity part-time, they think your brain has idled...when I was part-time I used to be passed over for courses and things, they thought if you had kids and you were part-time, you were not interested in progressing. That was very far from the truth. I remember being passed over for a course when I was pregnant, and I asked if I could do it and he said no I couldn't because I was pregnant.'
>
> (N18, *working mother of two, senior nurse*)

> 'The girls who work full-time have more status, the less hours you do the less standing you seem to have. They look down on you. I am going to lose status with my colleagues when I reduce my hours.'
>
> (N8, *working mother of three, midwife*)

In summary, the majority of women returned to work part-time after having their child(ren) believing it allowed them the opportunity to combine the social and financial benefits of working, with the opportunity to be at home with their children whilst they were young. For many women, mainly those who were simply interested in the social aspects of getting out of the house and gaining adult stimulation, it was the perfect solution, combining 'the best of both worlds'. However, for those who looked at the financial side, although it provided the security of an additional regular income, it was often not enough to make a big difference to the family standard of living but yet too much to give up altogether. For those women wanting to maintain their careers, they felt almost stagnant in their position as they lost status at work, and were often unable to progress with promotion whilst they remained part-time. This supports Richards' research into inequality in the workplace which concluded that whether domestic commitments actually do interfere with the working life of an individual woman or not, such generalisations are made, and are used as justification for the poorer promotional aspects experienced by women (1988, 155). A report by Dex (1986) concluded that in fact childbirth is undoubtedly the biggest single cause of downward job mobility, the key factor being not so much having the baby, but the indirect effect of moving into part-time work afterwards. Part-time work therefore remains an option for women with children; however, they need to negotiate clearly with their employers what they wish to gain from their employment in order make sure they benefit.

Childcare

As has previously been discussed in chapter 3, childcare is often seen by both academics and the Government as *the* biggest barrier to women returning to work. In contrast with the National Childcare Strategy (discussed in chapter 3), the two biggest issues raised by the women to regarding formal childcare were cost and flexibility. Only very few women managed to cover their childcare relatively cheaply and they indicated that they felt 'very very lucky, I am lucky' (N13).   Childcare often consisted of a mixture of after-school clubs, childminders, nannies, friends, family and neighbours, only rarely would one source of childcare be used.

> 'It was a patchwork really because there were childminder days, playgroup days and working in the evenings to fill in the gaps.'
>
> (D13, *mother of two, works from home*)

Some women who did not use formal childcare, said that it was because they could not afford to,[3] and those who did, indicated that a large proportion of their wage was used to cover the cost. However, the problem with childcare is not just a financial one. As discussed above, some women did indicate that the cost of childcare was a barrier; however the majority of women had other reasons for not wanting to use it, such as wanting to be the carer themselves and not feeling comfortable leaving their children with someone else. An aspect of childcare which all women did comment on was the need for flexibility. As we move more towards a 24/7 economy and people are expected to work shifts, longer hours and unsociable hours, there is a demand for childcare to be more flexible and become more readily available outside of the normal 9-5 regime.

> 'I wanted someone who we could trust, we had asked around and I couldn't find anybody who was particularly suitable. I didn't like a couple of the nurseries we had looked at, the other difficulty is because of the shifts we work, so someone who is prepared to have a baby from 6am. The hospital had a crèche, but they started at 9am, so that was no good. Also finishing at 9pm, no one is prepared to look after them till that time at night, so it wouldn't work.'
>
> (N18, *working mother of two. Her and partner work opposite shift patterns to ensure someone is home with the children*)

Another example is provided by a respondent who is a (surgical) theatre manager and if the theatre list is running late, or a patient has not recovered from their operation, then she cannot leave work, she must stay with the patient. This has occasionally caused problems with the childminder, as at the last minute she has

---

[3] Other reasons for not using childcare given by women at home will be discussed in the next chapter.

had to call and say she will be late; however, she has a good relationship with her childminder and feels comfortable to be able to do this. Though this was not a common story, most women talked of the anguish and stress they experience in trying to ensure the children will be picked up by someone they know and trust, should the need arise. If possible, women would turn to family members to help out in times of need; however, it is the women who had no family in the area who found the inflexibility of childcare the biggest problem.

In order to get around the necessity for flexibility in childcare, some of the parents who were both in professional jobs (where they may regularly require flexibility), hired a nanny during the pre-school years.

> '[She] came in everyday, and I liked that option because she was coming into our home so if the children were ill, they were in their own home. They did not have to be bundled into the car and taken anywhere, and it meant that I did not have to take time out, because if they were at a nursery or a childminder's, you would have to keep them at home, so it had lots of advantages, although the cost was horrendous.'
> (D18, *working mother of two, partner is now at home full-time to look after children*)

> 'The childminding thing, you have got to be able to take your children somewhere and pick them up. Very often people want set hours, and we were looking for someone with a little more flexibility.'
> (D2, *working mother of two*)

All the women who hired a permanent individual childminder for their children talked of the problems of finding someone to fulfil the role to their satisfaction and also went through two or three nannies before finding the right one. However, once they found a nanny which they liked, all the women praised them highly and commented that they would not have been able to fulfil both their dual roles of mother and worker without their help. Due to the nature of the relationship demanded between a mother and the nanny, in many cases the nanny has remained a close friend of the family, even being the child(ren)'s godparent or in one case the children were the bridesmaids at their nanny's wedding.

> '...she is like a grandmother really...they get Christmas presents, birthday presents, Easter eggs, if they go on holiday, they buy her presents, it is more than just a minder.'
> (N13, *working mother of two*)

In addition to cost and flexibility, women talked of the additional difficulties they faced with two or more children. All women with two children requiring childcare stipulated the necessity to keep them together. However, finding suitable, external affordable childcare with two places available can often be difficult. As an alternative, although employing a nanny for one child is very expensive, it became financially viable with two or more children.

Opinions on the availability of childcare were contrasting even within local areas; some women would tell you there was lots of choice, whereas others in a nearby street would indicate there were no suitable childminders in the area. Although there may be registered childminders in the area, whether or not a mother deems them suitable is another matter. Each parent has a different opinion on what is 'suitable', and many women gave me accounts of childminders they knew or had heard of who were definitely not suitable. However, in all probability the parents whose children were with those childminders were happy. Therefore, there is no easy answer for the Government who is trying to provide suitable childcare places for all children under five years old through the National Childcare Strategy.

Conclusion

This chapter has demonstrated that it is a complex combination of interdependencies which overlap to influence a woman's decision on whether or not she wants to, and is able to, (re)enter and negotiate the formal labour market after having children. The interdependencies and barriers for women (re)entering into the formal labour market can be considered in two categories; the practical barriers such as the lack of suitable childcare, lack of flexible working arrangements and lack of suitable employment, and the social barriers such as the presence of an informal male-dominant culture and predetermined societal expectations of woman as homemaker and male as breadwinner. In relation to the two case study areas, the working women of Neath Port Talbot faced more social barriers when negotiating the formal labour market in comparison with those from West Dorset who expressed more practical barriers. Once in employment the majority of women were there in order to gain from the social benefits that employment brings, such as adult interaction, having a sense of worth outside of the home, gaining confidence, new friendships and 'getting out of the house', rather than solely for the financial gain. However, all of the working women I spoke to in both Neath Port Talbot and West Dorset felt that if employers were more understanding of the needs of working mothers, and allowed them more flexibility to do their job, then more women would (re)enter the labour market after having children. In hindsight, when looking back at the decisions they had made about being a working mother, very few of the working women I spoke to said that they would change anything. In fact, those who indicated they would have done something differently, it was often that they would have gone back to work sooner as they felt the benefits outweighed the difficulties.

All too often government publications and some academic research argue that childcare is *the* barrier to women entering into formal employment. However, this chapter has outlined some of the other key interdependencies which women have to negotiate on a daily basis, indicating that other factors, such as predetermined expectations and their partner's opinion, have just as much, if not more, influence as whether or not childcare is available. The next chapter investigates whether the women who are at home with their child(ren) full-time encounter the same issues and considerations as the women who work. I shall investigate whether these women have 'chosen' to remain at home, or whether they truly wish to enter into the formal labour market, but feel unable to do so.

Chapter 7

Exploring the Interdependencies: Women and Home

Introduction

An overwhelming majority of the women who were at home full-time were there because they had chosen to leave the formal labour market in order to look after their children. This chapter explores the issues surrounding women and home by looking at the reasons women have for making the choice to be at home full-time, and what impact that has on the family, the children, and their future working career path. Almost all the women in this 'category' were in employment prior to having their first child. However, some women had already made the decision that once they had their child, they would stay at home; for others, the decision came later. The first part of this chapter looks at the issues raised by women who have made a proactive choice to be at home with their children by firstly revealing the reasons women feel that a mother should be at home with their child(ren), and secondly, I explore the thoughts that women, who have chosen to stay at home, have on childcare and why they are not happy to use it. The second part of this chapter looks at those women who may not necessarily be at home through choice, for although they may want to go out to work, they are not able to get a job which fits in with their domestic responsibilities. In relation to all the women who are at home with their children, I discuss the different attitudes which women have towards money, given that they are not directly contributing to the family income themselves. Additionally, I examine their home life by exploring the split of the domestic chores within the family household both for those women who are at home full-time with their children, and those women who are employed in the formal labour market. Finally I explore the women who could be perceived to cross boundaries between women at home and women at work. Firstly, there are those women who work from home, who argue that they get the best of both worlds by being able to negotiate their employment around their domestic responsibilities. Secondly, there are women who officially declare themselves as 'housewives', yet partake in 'cash-in-hand' employment to give them a little extra cash each week. Finally I shall look at those women who, after having a break from the formal labour market to look after their children, returned to college to further their education, either to update their skills, or to retrain for a new career.

Why Women Want to Stay at Home

Many of the women who were at home with their children full-time had made the proactive decision to do so, prior to the birth of their child(ren). The majority of women at home strongly believed that if you are going to have children, then it is your responsibility to look after them, and if you are going to 'pack them off to someone else every day' (D5), then there is no point in having them.

> 'I have always believed that if you are going to have a child, then you look after it.'
>
> (D5, *mother of four, housewife*)

> 'I just felt that I really wanted to have children and I wanted to devote my time to them especially whilst they were pre-school years. I felt that if I was having kids then I should be there to look after them.'
>
> (D8, *mother of four, thirteen years out of labour market*)

> 'I don't like the idea of people picking them up who they don't know. I want to be there when they get up and when they come home, and cook their tea...If they were ill, I would never ever leave them and go to work, even if it was just a flu or cold. Definitely not, no. I know when my children are ill, it is me they want, no one else.'
>
> (N1, *single mother of three, at home full-time*)

The women believed they should be at home with their child for a variety of reasons including being the one to influence the development of their personality, bringing them up to their own standards not other people's, being able to watch their first moments which can never be replaced and simply not being able to leave them with a stranger all day. All of these reason will be discussed in further detail throughout this chapter; however, at this point I wanted to highlight the fact that the majority of women I spoke to who were at home with their children full-time, did so because they believed it was the right thing to do.

Being at home with the children is not a decision which is always made in advance. Prior to the birth of their child, some mothers thought that they would return to work after their maternity leave; however after the birth they realised that they did not want to leave them.

> 'I was working full-time and was planning to go back, but from the moment I had him in my arms I decided not to... I can't see the point in having children if you can't look after them yourself.'
>
> (D15, *mother of two, self-employed accountant*)

For other mothers the decision came later. D2 had returned to work after maternity leave with her first child, and again after the birth of her second child. However, she did not return to work for long, and she soon realised that it was too much for her and she left work completely to look after her two children. D17 also returned to her previous employment in the family business where she had worked for twelve years. Although it was her mum who looked after her daughter for the few hours she was at work, D17 was not happy because she wanted to be a full-time mother to her children.

> 'I did go back to work when she was 6 months old, part-time whilst my mum had her. It was just 2-3 mornings a week just to help out dad. But in the end I just said no, I couldn't do it. I didn't want to do it, I just wanted to be at home and look after her.'
>
> (D17, *mother of three, full-time carer for disabled son*)

A few women discussed how, prior to having children, they were naïve as to the extent to which having a child would affect their lives.

> 'I think before you have children, you have no idea about what it all means. You think that you will be able to organise things beautifully and it will all slot into place and the child will go to a childminder and you will go back to the office and nothing will change. But actually I worked very long hours and it would have been completely impossible. I certainly couldn't have gone back and done the same job. I knew on day one, I remember the night he was born thinking, there is no way I could leave this child with anyone else. That never changed, I never felt any differently.'
>
> (D15, *mother of two, self-employed accountant*)

In contrast, one woman in Neath Port Talbot described how, when she got married, they decided that when they had children she would stay at home full-time to look after them because she believed that 'you miss all the first smiles, first teeth, everyone else gets them, you don't and you can't replace that' (N10). Immediately after they were married they started to make preparations for this by using only one wage to cover everything – mortgage and living costs – and then the woman consciously staggered giving up work before she became pregnant so that she could get used to being at home. These preparations were carried out so that when it came to giving up employment, the family would not notice the difference.

> 'I worked full-time for about 6 months after we got married and then for 6 months I job shared, so I worked 2 days one week and 3 days the other...We felt that way I would get used to being home, I became pregnant then, we had been married 18 months and I had a little boy.'
>
> (N10, *mother of two, housewife*)

Women who were at home with their children when they came home from school, talked of the importance of being there to listen to their child(ren)'s tales of the day as soon as they come home. Both working mothers and mothers at home acknowledged that if the child has been with a childminder, by the time that a parent picks them up, then the child has forgotten what has happened and how they feel about their day. One mother who is at home talked about how her son was being bullied at school, and she picked up on it straight away as when he came home from school he was quiet and subdued, not his usual self, and she therefore went up the school the next day to talk to the teachers. D5 argued that

> 'If he had lots of other people looking after him it could have taken a lot longer to notice.'
>
> (D5, *working mother of four, housewife*)

> 'I want to hear what they have been up to all day, by the time they come home from the childminder they are tired. You might hear about it the next day or the day after that.'
>
> (D1, *working mother of two*)

> 'I think they deserve their mum when they come in. It is like a sounding board for them, but if she was going to a childminders, she wouldn't be able to get it out of her system when she comes through the door, she would have to sit there and be polite. But now she comes in and moans about it all and I listen.'
>
> (N15, *mother of five, housewife*)

Although there are many benefits to being at home full-time in order to bring up your children, almost all of the women acknowledged that being at home full-time is not always easy: not only is it physically tiring continually chasing around young children, but also mentally, as mothers can get frustrated by the lack of adult conversation and not having an external focus other than the children.

> 'It does drive you mad sometimes when we do not go out for a walk or whatever, when you are stuck inside with her, but we survive.'
>
> (D14, *mother of three, housewife*)

> 'I must admit I was getting to the point where I was almost crawling up the walls because I found it very frustrating.'
>
> (D13, *mother of two, had 15 months out of labour market*)

> 'I did do nothing for about 3 months, but I was climbing the walls, I just couldn't stick it. I have to have something to look forward to, at work...something to go to.'
>
> (N12, *mother of two, works a few hours for family business, volunteers for Citizens Advice Bureau*)

> 'Then when he was 2, I found I was starting to vegetate basically. If I had gone to another coffee morning or a mother and toddler group, I would go mad.'
>
> (N14, *mother of three, two years out of labour market*)

Although women may have made the decision to be at home with their child because they felt that to be their sole carer was the right thing for them to do, there are also many other factors which influence a woman's decision to be at home full-time, including negative feelings about childcare or not actually being able to get a job to fit in with other domestic commitments. I shall address each of these issues in turn.

Thoughts on Childcare

Within both of the case study areas, there was a generational difference in opinions on childcare. The majority of women with older children (now aged about 15 years or over) expressed that when they had their children there was no childcare available and so there was no other option but to give up work and look after their children themselves.

> 'Crèches and nurseries weren't available then, it was just assumed that if you had children you would stay at home and look after them.'
> (N15, *mother of five, housewife*)

> 'There wasn't that much around then really, not round here anyway. There are lots of crèches and childminders now, but there wasn't any then, 24 years ago.'
> (N6, *mother of three, eighteen years out of labour market*)

In more recent years, as childcare options have grown and become more publicised, and more women are returning to work after having a child, opinions on using childcare amongst the younger generation of mothers have split. The majority of the women who were at home with their children had negative thoughts about leaving their children with childminders, for they believed that they were the best person to bring up their child, and would not want to leave them with someone else whilst they went out to work. Although it may be a registered childminder, a neighbour or a friend, some women were not comfortable with allowing someone else to influence their child's upbringing. In fact many of the women I spoke to had friends or knew of someone who was a childminder, and although they thought they did an excellent job with other people's children, they would still not trust them with their child(ren), for they believe that it is important for the mother to be with her child, particularly during the early years of the child's life as the child develops.

> 'We felt that if we made the decision to have a family, then we were the ones to look after it. It is your decision to have a family, it is not fair to oust them onto somebody else, you can only blame yourself then if things go wrong.'
> (N10, *mother of two, housewife*)

> 'I don't want my children to grow up with someone else's standards and ideas... Not that there is anything wrong with their standards at all, but when you have your own rules, for children it just wouldn't be fair because one minute they are being told one thing and the next another.'
> (D6, *mother of two, five years out of labour market*)

Women who believed that they should be at home with their children also commented that women who are at work when their children are young miss out on so much of their child's development. It is not only the first word and the first steps, which can never be replaced, but also the development of their child's personality.

> 'I think that you miss out on so much if you go to work, you miss all the first smiles, first teeth, everyone else gets them, you don't and you can't replace that... I think if you stay home you can concentrate on education, like he knew his colours before he went to school, he started recognising words before he went to school.'
>
> (N10, *mother of two, housewife*)

> 'I felt that firstly I had missed out with everything going on with Rhiannon. I would go down to my mother-in-law's and she would tell me things she had said, that she was crawling, walking and all the rest of it, and I didn't want to know, I wanted to see it for myself.'
>
> (N18, *mother of two, worked full-time with first child,*
> *now works part-time*)

However, it is not just for young children that mothers feel they should be at home, for some mothers who have been at home full-time and have subsequently returned to the formal labour market find it difficult to not be there as soon as their child(ren) come home from school, even though they are not needed as the children grow into teenagers. These mothers feel that it is still important to be there to help them with homework and to cope with the pressures of studying for exams.

> 'Although they are 14 and 16 I think that I should be here. I still feel that I should be here. They don't need me, but I feel I should be here. It is hard to explain.'
>
> (D6, *mother of two, five years out of labour market*)

Although many of the younger women who are/were at home with their child(ren) respect that childminders often do a good job, they would never feel comfortable leaving their child(ren) with any form of childcare, albeit a registered childminder, a neighbour, an after-school club or a friend. Many of the mothers felt that if you choose to have children, then it is the mother's responsibility to be there, not a job to be left to someone else. Although this was the opinion of the majority of mothers who were at home full-time, it was not the only opinion. Some of the mothers I spoke to would like to enter the formal labour market, but are unable to find 'suitable' employment.

Not Able to Get the Job I Want

Some women who were currently at home with their children expressed how they would love to get back into the formal labour market, to give them something to do outside the home, to get out of the house and to meet new people. However, the majority of these women wanted employment to fit into school times, as their priority was to be there for their children after school. They knew this would limit their opportunities and so many of them have given up proactively looking for employment. This was a situation particularly prevalent in Neath Port Talbot where many of the women I spoke to said that they would love to enter the formal labour market but jobs were not available which fit with the children.

> 'I would like one day, just to do something different in the week. I would, but I know I can't… I was thinking I could do something like, even a shop job, just for a couple of hours a day. Perhaps 11am-3pm, as long as I can get back for the children, to take them and fetch them.'
>
> (N1, *single mother of three*)

> 'I would like to work full-time, but I have no one to have the little one in the mornings. The afternoons are not too bad as either he is here or my mother comes down. But the mornings, there isn't anyone. I was offered a job this week in the factory, 8-4, but it is that hour between 8 and 9am.'
>
> (N5, *mother of three*)

> 'At times when I am home, like now, with only her for company I don't see anyone, I think I would love to be out somewhere just to have a few different work colleagues who can have something different to talk about when you get home in the evenings.'
>
> (D14, *mother of three, housewife*)

Having been out of the labour market for a while, one mother described how she wanted to train as a midwife, and completed a GNVQ in Advance Social Care. However, the specialist training involved time 'on call' with a community midwife when she would be expected to attend at any time of the day or night. As a single parent, this would mean leaving her children on their own, something she was not comfortable with doing, and so she had to give up the course and look for alternative employment. This was not an unusual situation in Neath Port Talbot, as many of the women felt confident that they would be able to retrain and start a completely new career. In fact, lots of women in the Neath Port Talbot region did go back to college to do a variety of courses including cake making, sociology, law, costume design, catering and book keeping as a way of gaining new skills before (re)entering the workplace. I shall discuss this in more detail later in the chapter.

The lack of 'suitable' employment is not the only barrier for those women wanting to (re)enter the formal labour market. The cost of (re)entering the labour market or changing job was also a barrier to some women; for example, the cost of retraining or updating skills or perhaps needing to run two cars. One lady in Neath Port Talbot had been a night sitter in a residential school for many years, and was tired of working nights and not getting the sleep she needed as she was then looking after her children during the day. Over the years, she had also lost the social benefits she used to gain from work, such as the adult interaction, as others had been made redundant and she was the only one there at nights. Although she was incredibly unhappy where she was, N3 did not think that it would be possible for her to change jobs due the potential expense of travel and having to buy new clothes for office work.

> 'People say why don't I change jobs, do something that, but I think if I had another job, firstly it is travelling expenses whereas I get a lift into work and walk home in the morning. If I go and do office work, it would be clothes to wear, it is a big expense if you are on a tight budget, whereas up there you can go in jeans, so all in all that is why I have stayed there.'
>
> (N3, *mother of two, night sitter in residential school*)

A similar concern with money was expressed by D10, who had previously worked as a senior nurse in West Dorset but, after the death of her partner, was no longer able to do the shift work the job required. For many years D10 has been on benefits and worked as a shop assistant, but is now looking to return to nursing. Having been out of the profession for more than five years, it is necessary for her to do a Return to Practice course to get up to date with her skills. Although the course is free, it is unpaid and will require full-time commitment, and D10 faces financial problems with the cost of transport to and from the college and as a consequence is going to have to continue to work in the shop for 16 hours a week in addition to attending college full-time.

> 'When I start my return to practice course which is next Monday, I get paid as a grade B, but not my study days. I have two study days on Monday and Tuesday next week where I have got to get to Bournemouth for 9am, and I don't get paid for those days. I have to go to the uni for classes. Between then and my next study days I have got to do at least 3 days on the ward, but NHS pay a month in arrears for Bank work, so anything I earn in May, I will get at the end of June. So, unless I continue working in the shop I have got no cash in my pocket to get to Bournemouth.'
>
> (D10, *mother of two*)

This section highlights the fact that, although for the majority of women in Neath Port Talbot and West Dorset, being at home was a proactive decision, some women indicated that they would like to join the formal labour market. However it must be stressed that their priority remained their child(ren) and any formal employment would have to fit in and make no impact on the existing daily routine, therefore reducing the prospect of them (re)entering the formal labour market.

Although they indicated that they would like to enter employment, in reality, these women retain the choice to be home with their children as their priority.

Domestic Chores

I would just like to briefly discuss some of the thoughts that the women had on the domestic duties within the household, for although not necessarily a direct influencing factor on women's decision about whether to work or not work, for as argued by Mattingly, Hanson and Pratt (1998) the addition of a child to the family substantially increases the domestic responsibilities within the household. The uneven split of domestic responsibilities is an issue which most women, both women in the formal labour market and women at home, raised as something they were not happy with, but had over time come to accept.

Women who are at home with the children argue that looking after children is a full-time job; however, in addition to being a full-time mother, they are also carrying out almost all of the domestic duties. Many of the women said that they would not expect their partner to come home after a day at work and then cook the dinner instead, the women see the domestic duties as part of their role of being in the home.

> 'I don't expect him to come home and then have to clean or cook his own tea. He has been working hard all day, and I have been here. I just get on with it, it is just one of those things you do.'
>
> (D14, *mother of three, housewife*)

> 'When he is here, if something needs doing then he will do it, it is just because he is physically not here.'
>
> (D15, *mother of two, self-employed accountant*)

However, the majority of women who were in the formal labour market undertook the 'double-shift' undertaken by many mothers who have to do the housework after getting home from work (Freeman 1982) even if the partner came home first. For as Sichterman (1988) argues, the majority of women discover that when they take on a job they could not really count on help with the housework from their husbands, as described by D6.

> 'I used to work overtime and get in around 8 o'clock, my husband would have been in since 5 o'clock, but I would then still make dinner. He was supportive, but he relied on me to tell him, and he would do it, he would never object.'
>
> (D6, *working mother of two*)

This was a common situation with working mothers as Luxton's (1990) research into the gendered division of labour in the home discovered. D6 described how sometimes her partner would perhaps clean up the kitchen ready so that when the she came home she was able to get on with cooking the family meal. Many of the women, both those who were in employment and those at home, commented that if they were to ask their partner to do something, then he would usually do it, but he would rarely initiate domestic duties himself.

> 'If I leave him instructions for something to be done, then it would be done, but the pile of ironing would sit there.'
>
> (N18, *working mother of two*)

> 'He will do some ironing, and peg out if I ask him. If I ask him to do something, he will do it, but generally I do most of it. Men don't like to think that they are able to do anything do they!'
>
> (N5, *working mother of three*)

Many of the women were also keen to point out that if their partner did cook a meal, or run the hoover through the house, then they would make sure that the woman knew they had done it by telling them and reminding them repeatedly that they had done it. This supports Benn's (1999) argument that men's help in the home, is just that, help, rather than an acceptance of equal responsibility, as N7 describes:

> 'He will make a meal once a month and then I won't hear the last of it. Sometimes I wonder why I bother. Perhaps he will iron something and that is it, you could swear he had done a basket full of ironing every night for a year, but he has just ironed one shirt.'
>
> (N7, *working mother of three*)

This was a common story for all women, as they described how it had become expected that they were responsible for the domestic duties, and although they did not like it, they appeared to accept that it is easier for them to just get on with it than to go round and clean up after them.

> 'I have always done it all. He once washed in 1989, and it shrunk something so he has never done it since. He will do the odd thing. He does cook, he will do a meal at the weekend, and he will leave a lot of pots and pans out and I have to clean up the mess.'
>
> (D5, *mother of four, housewife*)

There were only a few women who felt that the divide of domestic duties was equal, and these were all women who were working, and their partners had had to accept responsibility when they (re)entered the labour market. D18 described how as a solicitor in private practice, she needed to put in long hours in order to succeed. As she is the breadwinner for the family, her partner has now taken over responsibility for the two children, ensuring that they get to and from school, and preparing the evening meal; however she remains in charge of the housework such as cleaning the bath!

'If you are going to succeed in private practice you have to put the hours in, so therefore the hours are shifting. So Nick does all the running around and he cooks the meals and works when they are at school, but I am responsible for the laundry. I used to be extremely house proud, very much so, but that had to go…If he waited for me to come home and cook, it was a waste of time, and so he said he would cook. In fact he has now developed an interest in cooking. We don't just have shepherd's pie any more, we have monk fish on a bed of steamed vegetables! So now he will like to try out Rick Stein and Gary Rhodes recipes… He does see the need to clean the bath, but can't do it, but then that is life.'

(D18, *working mother of two*)

The split of domestic responsibilities remains a highly contentious issue within the household as the majority of women felt that their domestic work was undervalued. This supports Castells (1978) who argues that use of the concept 'consumption' implies a failure to recognise that the work that women do in the home, as domestic labour, is work. Although it may not directly influence a woman's decision regarding whether or not to enter the labour market, as discussed in chapter 6, the willingness of a partner to share responsibility for domestic chores can limit the type of employment a woman can enter into (see Pratt 1993).

Attitudes Towards Money

Many of the women at home with their children full-time had made that decision regardless of the financial pressure that would place on the family. It was more important for them to be at home with the children than to have the materialistic 'extras' that a second wage would provide for the family. Many women, who had proactively made the decision that they wanted to be at home with their children, did discuss the fact that by doing so, they had had to make sacrifices. For example due to living on one wage, women at home full-time were unable to afford the luxuries which other families were able to enjoy, such as the family holidays abroad, expensive Christmases or birthday parties; however, for these women it was a small price to pay in exchange for the pleasure and benefits of being home with their children full-time.

'Christmas they never get toys off [the grandparents] they will have clothes, we never have Easter eggs from the grandparents it is clothes or whatever … They can't have things just because they want them, they have got to wait till Christmas or Birthdays or whatever.'

(N10, *mother of two, housewife*)

'I have had money, so I have given the kids pocket money and things. Christmas was great. Since we have been down here we have had really skimpy Christmases we really did and this year we didn't at all. All my wages went on Christmas and that was really good as I was not feeling guilty because I was not taking money from the farm. At Easter we went out and I brought us ice-creams.'

(D7, *returning to work after twelve years out of formal labour market*)

The majority of women who were at home with their children acknowledged that at times they had struggled to make ends meet financially, sometimes putting the family in financial insecurity. It was at times like these that some women had considered the possibility of going out to work themselves, but realistically, it was not what they wanted and it would only have happened as a last resort. In fact, although many considered the option, not one of the women I spoke to in either Neath Port Talbot or West Dorset returned to work solely for financial reasons. For other women, the idea of going out to work and leaving the children with childminders was never considered a realistic option for they wanted to be at home with their children, and so if they hit financial difficulties, then they just got on with what they had and made it work; being at home with their children was more important.

'At the moment we are struggling a bit, but you just put up with it. You pull everything in, tighten your strings and get on with it. I would rather do that and still be at home with my children.'

(D17, *mother of three, housewife*)

'Once I knew I was expecting, I wanted to give up work. I knew we would only be on one wage, but the way I looked at it, there was plenty of time to go back, I just wanted to spend every minute with her. It was a struggle, I will be honest, but we had bare walls for 10 years in the hall, but I was quite happy because I knew it would come, there was plenty of time.'

(N3, *mother of two, had 3½ years out of labour market*)

'Definitely, never considered going back to work. It has been a struggle financially, but you have got to make the decision that if you want anything you save for it. It is worth it when you see your child develop... It is a struggle living on the one wage, Andrew was on a 3 day week at the beginning of the year. How we managed I don't know.'

(N10, *mother of two, housewife*)

'It was considered. It is always there at the back of your mind, it is not something I would have enjoyed. We did discuss it, but who would look after the children? ... No, it just wasn't worth it.'

(D17, *mother of three, housewife*)

Within both Neath Port Talbot and West Dorset, there were very mixed reactions from women at home regarding the family finances and how comfortable they felt with spending money, considering that they were not directly contributing to the family income. Some women felt completely at ease with buying whatever they wanted or needed, as they had the attitude that they were married and whatever was his was theirs as well, as they were in a partnership. Although a mother at home may not be bringing home a wage, they argued they were still working hard doing a 24 hour job, and so if they wanted something, they could have it.

> 'They think if you are staying home you are not fulfilling yourself. Well I think motherhood in itself is a job and you don't realise what is involved until you are actually doing it. If the mother had to be paid for staying home, then we wouldn't get what we deserve for what we do. You are a mother, a cook, a cleaner, nurse, chauffeur, everything rolled into one. You can't put a wage on that.'
>
> (N10, *mother of two, home full-time*)

However, some women who were not bringing in an income to the family felt very uncomfortable with spending money which they did not feel was theirs, and felt they had to ask before they wanted to buy anything, and only then if it was absolutely necessary.

> 'I wouldn't spend any money when I first gave up work as it wasn't my money.'
>
> (D2, *mother of two, had one year out of labour market*)

> 'When I wasn't actually working, I had to ask for pocket money. I had no money of my own, I had to ask for pocket money. I found that quite galling and he was quite shirty about it. I felt I had to account for every penny we were using.'
>
> (D16, *mother of three, had 18 months out of labour market*)

Contrary to what may be expected, money was *never* the sole influence on a woman's decision regarding whether to work or not work. The majority of women at home were there because they wanted to be, and if money was short, they simply tightened their budget and got on with it: returning to work would have been the last resort.

The next section of this chapter looks at two groups of women who cross the boundary between women at work and women at home, for they are both employed and home with their children when they need to be. Firstly, I shall look at women who are employed within the formal labour market but work from home, so if and when they are needed, they are available to their child(ren). Secondly, I shall look at women who are employed but not in the formal labour market, rather they undertake cash-in-hand employment that is flexible enough to be organised around their domestic responsibilities.

Working at Home

Some of the women were in employment that enabled them to work at home, usually self-employed. These women spoke of the flexibility working from home gave them, as they were able to organise their working hours around their child(ren)'s demands. Often this would mean working for a few hours whilst the children were at school and then again in the evening after the children had gone to bed.

> 'I had to do it at night when the children were in bed, and at one point my sister and I did it together. At some points it got so big we were working through the night and it was horrendous.'
>
> (N9, *started a sewing business at home*)

> 'Working from home, I have never paid for childcare arrangements at all. It is the most straightforward thing in the world, especially with half terms and holidays and things, but I work in the evenings to make up the time, and my work is sufficiently flexible that I can do that.'
>
> (D15, *mother of two, self-employed accountant*)

D15 also sometimes found it could be difficult to work at home with children around, as clients would telephone at any time of the day and need her undivided attention. D15 explained that when the children were younger, there were times when they would be noisy and play up in the background; however she has brought them up to understand the meaning of value, and that in order to have the nice things like new trainers and holidays, then she must be able to work. The children therefore understand that if mum is working, then they must behave themselves and be quiet.

> 'I am available to my clients all the time, and the children know that if I have a client on the phone, they have to be quiet and that there are other priorities other than them.'
>
> (D15, *mother of two, self-employed accountant*)

Alternatively to organising the working hours around the demands of the children, D13 works from home as a freelance computer consultant, and hired a nanny full-time to be with the children in the home whilst she worked in the office at the back of the house. D13 described how it could well have been hard as she could hear the children screaming and creating havoc in the background and she was sometimes tempted to come away from her work to watch the children, even though she had compete faith in the nanny. The children would also come and bang on the door saying 'mummy, mummy, mummy', which was difficult to ignore. However, D13 believed that in order to succeed, she had to have good discipline and gave herself set working hours which she stuck to.

'I do not find it a problem to shut the door and go away when it is time
to work. I do not worry if someone may have spilt cereal on the table.
That will get cleared up at 5:30pm when I have finished.'

(D13, *mother of two, freelance computer consultant*)

Although the women who worked at home believed home-working to be a great
way of combining the best of working with being able to be at home with the
children, some of the women felt that other people, such as friends, family and
neighbours, did not appreciate that although they were working at home they still
had a full-time job commitment. The women felt that they were undervalued by
others because they did not physically leave the house to go out to work.

'I don't think that being at home is valued as a proper job. Working
from home is viewed differently from going outside to work.'

(D15, *self-employed accountant*)

One woman explained how her family are continually calling her asking her to run
errands and make phone calls for them. They argue they do not have time to do
them because they are out at work, even though D19 has a full-time job as a
farmer.

'They think that you are at home doing nothing and they are doing you a
favour giving you something to do. Because she doesn't go out to work
9-5 sort of thing, they think she is at home and can do it.'

(*partner of* D19)

When I asked D19 what her employment responsibilities were, she replied;

'It varies every hour of every day because my parents have three
cottages down at the bay which are summer let so I deal with all the
bookings and telephone calls, paperwork from that, I even clean for
them sometimes. I am the local rector's secretary and I deal with that.
At the moment that seems to be daily as he has lots of things planned. I
do paperwork for my parents, this farm paperwork. Some paperwork
for my friends' two sons who are self-employed. Guides and Brownies,
their paperwork.'

(D19, *primarily a farmer*)

Similarly, when you look at the average day of the other women who work at
home, because they do not have travelling time and are not distracted by chats with
colleagues and coffee breaks, they often manage to work harder at home and be
more efficient with their time.

'Working from home you get a lot more done as you are not hanging
around making cups of coffee and swopping jokes. If they pay me for
7.5 hours a day then I was actually working for 7.5 hours.'

(D13, *mother of two, freelance computer consultant*)

> 'I brought extra work home because I was getting through it quicker at home.'
>> (D9, *mother of three, packaging for a pharmaceutical company*)

The women I spoke to were highly complimentary of working from home, confirming Harrison's (1983) argument that it can suit women who wish to stay at home with small children, and women who dislike the discipline and time keeping of external work and wish to work at their own pace. However, it must also be remembered that it can be very isolating for women, who may end up not leaving the house from one day to the next, and according to Harrison (1983), home-working is seldom properly paid. The undervaluing of employment at home by others can lead to higher expectations and demands being placed upon a woman, and working from home can require strict discipline and routine in order to be able to sit down and work without distraction. However, as demonstrated above, once a routine has been established, often working at home can be more productive in addition to enabling a woman to be there with her child(ren) when they are home from school.

Cash-in-hand Employment

Not all the women at home were dependent solely on their partner's income or benefits; some of the women were involved in 'cash-in-hand' employment such as cleaning, caring or helping out at a children's club, which they were able to do whilst their child(ren) was in school. The most popular type of this employment was cleaning for a private individual, just a few hours a week; however one lady appeared to have three cash-in-hand jobs going at once in addition to her family benefits.

> 'I helped out with children's gymnastics at the leisure centre a couple of mornings and one afternoon a week. The shop work was added in as well and fitted around the gym nicely. It was just a bit here and there in the job to start with. Then somehow a friend said to me about the Body Shop home parties and I started doing those as well. I did those from May to November, but by the end of the year I was totally blown out...So I gave up the shop because I said I can't do everything. That was in October, but then in the January, Nick, my partner and I had a car accident, I had really bad whiplash, so I had to give up the gym, so I had given up everything. I had a knock on my door, "hello, we have heard you have been doing things you shouldn't have been doing", it was Social Security...I had my leisure centre outfit on, and they know I was doing that. He said he was not interested in what I was going to be doing, but interested in what I am doing now. I said, "well as you can see I am at the leisure centre" which I declare, so he said that was fine. He asked if had been doing anything else, I said yes I had been doing Body Shop

parties, but I had stopped. He asked if I was going to be doing anymore, I said no, so he said okay and that was it.'

(D10, *working mother of two*)

Although it can be argued that D10 should have been penalised as she was working at the same time as claiming benefits, this was the same for all women undertaking cash-in-hand employment. For the women who were married and shared their partner's income, they described how the cash-in-hand employment gave them a little of their own cash each week so that they did not have to ask their partner for 'money for silly everyday things' (D14). However, the majority of the women who were involved in cash-in-hand employment were single parents living on benefits such as the Working Family Tax Credit (WFTC) or Family Credit. The women argued that they really struggled to survive on just the benefits alone and the little extra they gained from unofficial employment made a big difference to them by giving them cash in their purse each week. Although WFTC is set up to encourage women to work, D9 actually found that there was no incentive to work harder by doing more hours (for example overtime) to increase your salary, for WFTC penalised you by taking away the extra money which is why she took on cash-in-hand employment, so it would not affect her benefits.

'Really for a long time I was just doing anything I could find. So for about three years I had not been out in the working environment because the jobs I was doing were private, I did a lot of cleaning privately, I did a lot of care work privately. I then decided that I needed to get back into the working environment...so I got a job packaging homeopathic tablets. All the other jobs I was doing were on the side really, cash-in-hand or whatever.'

(D9, *mother of three*)

This is a common story for all the women involved in cash-in-hand employment.

'Due to the benefits, you can only earn up to £15 a week before it affects the benefits. That is why I clean for a lady twice a week, and that is cash-in-hand.'

(N12, *mother of two*)

'The cleaning job at the nightclub which I got as my sister and brother-in-law work up there, that was cash-in-hand. I did another cleaning job briefly for a lady, that was another private arrangement.'

(D14, *mother of three, housewife*)

One of the benefits of the cash-in-hand employment which the women undertook was its flexibility. Almost all the jobs which the women undertook were flexible enough to be able to fit it around their domestic commitments as N12 describes:

'With the cleaning, the children come first, I can go and do it another day.'

(N12, *mother of two*)

Although cash-in-hand employment is illegal and these women face the risk of being prosecuted, for them, the risk is obviously worth it. The opportunity to be able to earn a little extra independent income each week to put 'cash in the purse' in addition to the minimal benefits or family credit provided by the Government, enables these women to buy small luxuries once in a while, either for themselves or for their children. As D9 described, the way in which the current WFTC operates, any extra money earned in the formal labour market is taken away by the Government, forcing women to turn to informal employment to boost their income.

Returning to Education

As discussed earlier in this chapter, even those women who had proactively chosen to be at home with their child(ren), often found that, after the children were going to school during the day, they started to feel the frustration of not having anything to do other than housework, and so many women returned to education. Some women took the opportunity to catch up on education which they felt they missed out on before they got married and had children; others found that being responsible for, and bringing up young child(ren), gave them the confidence to rethink their career options and used the opportunity to retrain into a new career because they were fed up with doing low-skilled, low-paid labour as they had before having children.

> 'I chose to do English because I enjoy it, Maths because you have to.
> They are not what I would choose to do, but that is the way the world is.
> I don't know though, I want to do more than skivy catering jobs, but I
> don't know what.'
> (D8, *mother of four, thirteen years out of labour market*)

Some women did college courses simply to give them something different to do, others wanted to gain official qualifications. Some women started by signing on to one or two evening classes so that they were able to go when their partner would be home with the children, others managed to do courses which fitted in with school times. Whatever the reason for the woman returning to education, they were all able to list more than one benefit they gained from it, including learning new skills, gaining more confidence and making new friends. As discussed in chapter 6, these are the same social benefits which those women who were in the formal labour market gained from going out to work.

Returning to education was particularly popular amongst women who had had a few years out of the formal labour market to be with their children. Many of the women found the need to update their skills or to learn new skills in order to be able to (re)enter the labour market, as for many women, computers had only just been introduced when they left the labour market, but they knew that they would have to become computer literate if they were to return to similar employment in the formal labour market. Many women said they thoroughly enjoyed the experience of looking after children full-time and so decided to gain formal qualifications to enable them to continue to work with children in the future. N4 described how she used to help out in the nursery school which her youngest child attended, until the point when she was at the school more often than she was at home. The manager suggested to her that she got qualified so that she could be paid for her time as up, until that point, all her work was voluntary work. Consequently N4 then returned to college and completed a GNVQ in Advanced Social Care and subsequently managed to get a job as a deputy manager in Neath College nursery.

> 'A job came up in Neath College where I had done my training, as a deputy manager there. I got that and was there for a year and a half, and the manager's job became available....I managed 3 day nurseries then through Neath Port Talbot College, and then the other job which came up after 6 months was the day care registration and inspection officer.'
>
> (N4, *mother of three, Day Care Registration and Inspection Officer*)

In general, the women found returning to education an easy transition as college courses often fit into school times, and they share the same holidays so there is no need to worry about childcare. The only time that childcare became an issue was with school inset days, when either the mother would simply stay at home for the day; alternatively, many of the women took their child(ren) into college with them. The women described how the lecturers had no problems with this as many of the courses were geared towards mature students, and other mothers would do the same.

> 'I found my children were getting older, so I needed to look again at a career. So I made the big decision then to go back to college, and I did my NNEB childcare diploma which I did full-time. But again, all my children were in school and being college-based, I was home in the holidays so I still managed to work evenings over here to supplement my college.'
>
> (N14, *mother of three, two years out of labour market*)

> 'I spent about 5 years in college learning various courses, some were part-time day and evening courses and others were just evening courses. Catherine was small so I was home with her and my husband would look after her when I went to college in the evenings.'
>
> (N2, *mother of one, dinner lady*)

Interestingly, many women appeared to get addicted to doing college courses. Once they had completed one course, enjoyed it and passed, they wanted to sign up for another course. Like others, N16 spent many years doing a variety of courses including sociology, psychology, law, catering, hairdressing, cake decorating and Chinese cooking.

'Once I enjoyed that sociology that was me gone then, I was so focused
on wanting one after another. I didn't know what the hell O levels and
A levels were going to do for me after because I had no focus of what I
wanted to do, I think I was just proud at passing it.'
                    (N16, *mother of two, had eleven years out of labour market*)

Initially N16 signed up on a course because she wanted something to do whilst
the children were in school: however, passing the courses gave her a sense of
achievement and boosted her confidence enough that after completing a catering
course she went on to try a few business ventures of her own, such as opening up
a pizza delivery company. Now, N16 has settled into hairdressing and is
currently employed at a local salon where she is learning more skills; however
this will not last long, as she has her sights set on her own salon in a couple of
years.

   However, returning to college and trying to study was not easy for everyone.
A few women found that even though college hours were during school time,
demands at home meant that sometimes their study had to come second as
described below by N9, who returned to college for two years to do a BTEC
National Diploma in Social Care, after eleven years out of the formal labour
market.

'Because the children were so young, they took priority, so if it meant
that I just got a pass for a piece of work, then so be it, I got a pass. They
were more important if they had something going on...I resigned to the
fact that all I needed to do was pass the course.'
                                                      (N9, *working mother of two*)

Previously, N9 had worked in the local factories; however, by retraining,
passing the course, and entering into a professional occupation, she is now
head of a team of paediatric occupational therapists working for Social
Services. Returning to education was particularly popular in Neath Port
Talbot, as many women had left school without any formal qualifications due
to the expectation that they would leave work as soon as they had children, and
so they never thought they would need them. Also, due to the lack of
availability of 'suitable' employment, more women in Neath Port Talbot used
the college courses as a way of accessing the social benefits which others were
gaining from their employment, such as getting out of the house, meeting new
people, and intellectual stimulation. This is demonstrated in the way in which
women continually signed up for a variety of courses, simply because they
enjoyed doing them.

   One woman (N1) had to give up her course completely because the
childminder she had for her children repeatedly let her down, and so it was
impossible for her to complete the course. The majority of women found
attendance at college an easy transition as the hours fitted in with the children and
it did not therefore disrupt the already established routine within the home. It may
be worth noting that N1's three children were all pre-school age at this time, yet all

the other women I spoke to who returned to education, had done so after their children had gone to school full-time. Many of the women I spoke to felt that they had not achieved to the best of their ability whilst at school, and having responsibility for children gave them the confidence to return to education, with the majority entering into professional employment as a result.

Conclusion

This chapter has explored some of the key interdependencies which women have to negotiate within the home, in order to fulfil their 'choice' to be at home full-time with their children. I also looked at the sacrifices they have had to make in order to do so, and the way in which women have come to accept their dual role, of domestic duties and employment in the formal labour market. It is fair to say that the majority of women who are at home full-time were there because they had chosen to be – although many women indicated that they would love to (re)enter the formal labour market, but they were unable to do so due to the restrictions this would demand. For example any employment would have to completely fit in with the already established daily routine, not interfering with their ability to complete their domestic role or stop them from being a full-time mother to their children when they were not in school. All these mothers believed that they were the best, and only, person who should bring up their children and were willing to take the sacrifices of lower household income and place a personal career on hold. However, as discussed in chapter 6, many of the women who were at home with their children felt that they were undervalued for being 'just a mum', and argued that being a full-time mum was more demanding than a full-time job could ever be.

In this and the previous chapter, I have outlined and demonstrated the complex network of interdependencies which women have to negotiate on a daily basis. It is a combination of both practical and social factors which influence a woman's decision regarding (re)entering the labour market once she has had children. Those are often locally determined and are not reflected in national level data or national analysis. The next chapter will discuss the implications of the research for future agendas in the theory and practice of the labour market.

Chapter 8

Making Sense of Place

Introduction

I have argued that spatial variation exists within national labour markets, which has implications for processes and practices at the local level. More specifically, chapter 4 explored how nationally aggregated data conceals important variations at the regional and local level. After acknowledging the work of Peck (1996a) and others (see Bauder 2001, Kreckle 1980, Odland and Ellis 1998, Granovetter and Tilly 1988), that labour market participation is profoundly shaped by the spatial constraints women face in negotiating both their home and work lives, I have argued that a full understanding of the complex system of interdependencies which underlie women's daily lives can only come from local level research. Chapter 4 established that there is no straightforward connection between the status of the local economy and women's labour market participation, and it therefore argued that local social and cultural processes must be explored in addition to economic ones if we are to fully understand the geographic variation of women's decision-making process with regard to their participation in the formal labour market. Chapters 6 and 7 gave accounts of the findings of semi-structured interviews with women in Neath Port Talbot and West Dorset about the decisions they had made with regard to entering or not entering the labour market once they had children. The two chapters highlighted the complex network of interdependencies, which affect this decision.

In this chapter, I draw together the empirical work and situate my research findings in the context of contemporary geographical and policy debates. I have engaged with both segmentation theory, which aims to provide an understanding of people's varying positions within the labour market, and the locality debate which aims to provide an explanation for the variation in local labour market activities. Together both of these sets of literature are helpful in providing a further understanding of, and explanation for, women's labour market activity. This chapter will firstly revisit fourth generation segmentation theory (see chapter 2) and, in light of my research findings, discuss the way in which the theory has to develop if it is to provide a fuller understanding of the geographies of labour market participation. Secondly, I shall also re-examine the development of the locality debate, and show how an acknowledgement of the spatial variation which exists between and within localities enables us to further appreciate that labour market processes operate in different ways in different places. It is the interlocking combination of local social, economic and cultural processes which influences an individual's decision making regarding whether or not to (re)enter the formal labour market. Thirdly I would like to use those two different academic literatures (segmentation theory and the locality debate), to draw out the key findings from my empirical research and consider the implications they might carry for public policy in this area.

Revisiting Segmentation Theory

As discussed in chapter 2, the development of labour market segmentation theory has seen a move from the simplistic dualist labour market, to a more complex understanding of the interlocking economic, cultural and social processes, which combine to influence, in this case, women's labour market participation. I shall now briefly revisit segmentation theory, summarising its development and highlighting where the gaps are in providing an explanation for women's labour market participation.

First generation segmentation theory recognised that men and women occupy different sectors of the labour market due to woman's second role, which she performs within the home. Parallel with the development of segmentation theory, Massey and Meegan (1982) recognised that women can become trapped within the most unstable sectors of the labour market due, amongst other things, to the perception that their priority is their home life and the assumption that they will not be interested in progressing their career. According to Hanson and Pratt (1995) this supposition evolved through an amalgamation of women's tendency to organise their paid employment around the schedules of their husbands and children, combined with the sexist practices of male employers and employees. As I argued in chapter 2, second generation segmentation theory divided the labour market into primary and secondary sectors, and identified the fact that men were more generally placed in the primary sector with higher wages, advanced working conditions and strong unionisation, and women were primarily placed in the secondary sector with low wages, little opportunity for promotion and poor working conditions. This maintains a hierarchy in the labour market, and makes it difficult to move between sectors, therefore trapping women at the lower end of the labour market. Third generation segmentation theory began to explore the fundamental dynamics of the labour market, and labour market theorists began to understand that an explanation for labour market participation should not solely be drawn from studying what the labour market had to offer, but was also affected by the process of social reproduction, including factors stemming from within the individual household. These segmentation theorists became aware that the institution of the family serves to structure the supply side of the labour market in the allocation of domestic responsibilities, and a full appreciation of women's position in the labour market could only be gained if the two spheres, those of home and work, were both considered simultaneously. There was an understanding that economic, social and political forces combine in determining how economies develop (Wilkinson 1983, 413), and so a multicausal explanation was considered necessary.

In a complete turnaround from first generation segmentation theory, segmentation theorists have now come to realise that the family and other local social processes, as well as the type of employment available, are key factors influencing a woman's desire and ability to (re)enter into the formal labour market. It is the recognition of the spatial variation of local social processes, and the impact of the local processes on individual lives and choices that represents an important stage in the development of segmentation theory, and lays the basis for a fourth generation. There is now an understanding that the supply of labour is not only governed simply by market forces, but by demographic factors and social norms concerning the participation of different groups in wage labour. Evidence has been provided on the difference in the status of the economy in the two case study areas, which also represent the extremes of women's economic activity in the UK. However, the results from my questionnaire indicate that there are no significant differences in the importance the women placed on finance. Almost all women indicated that more money would be helpful, but it was not the most pressing issue for mothers. In order to develop segmentation theory further Bauder argues that the task for the labour market theorist is to gain a better understanding of the complex and interlocking relationships between demand and supply-side processes. He believes that fourth generation segmentation theorists should focus on place as a constitutive force to do this (2001, 47). This book has attempted to contribute to this development through focusing on two local case study areas as a way of observing the complex interlocking processes which determine whether or not a woman is able to (re)enter the formal labour market once she has had children. Bauder (2001) also claims that fourth generation segmentation theory recognises that labour market segmentation is partially a locally constituted process. What I have sought to do is to unpick and analyse the interdependencies which come together in different places to provide a locally sensitive understanding of labour market segmentation.

Extending Segmentation and Space Through Locality

Chapter 4 noted how authors within the locality debate have sought to analyse locally derived social processes as well as economic factors. In an analysis of 'locality as a process', chapter 4 argued that it is the connection of many interlocking systems of difference that creates an individual locality. In light of the findings from research in two localities with very different local economies (see chapters 5, 6 and 7), I would like to briefly revisit the locality debate and consider the way in which it can further inform labour market segmentation theory, if it is to help us to gain a full understanding of women's labour market activity.

The Difference that Space Makes: Locally Predetermined Societal Expectations

In a re-examination of the locality debate it is possible to see that some authors have highlighted the existence and importance of predetermined societal expectations (see Morris 1991, Cox and Mair 1991), local culture (see Duncan and Savage 1991, Jackson 1991) and the pressure for 'appropriate gender role behaviour' (Morris 1991, 167). Morris believes that these expectations and predetermined roles are often generated through the economic history of the area, as demonstrated by Dennis et al. (1956), who provide a study of a coalmining area in Yorkshire during the 1950s. Here the traditional sexual division of labour of male breadwinner, female homemaker was supported not only by the male-dominated employment, but also by the local culture of working men's clubs, peer group pressure, attitudes towards, and the type of employment available to, women in the area.

> 'The pure economic fact of man's being the breadwinner for his family is reinforced by the custom of family life, the division of responsibility and duties in the household, and the growth of an institutional life and an ideology which accentuate the confinement of the mother to the home...given the basic fact of a certain economic-social framework, family life and the accepted division between the sexes can build up a set of mores and ideas with an intrinsic force in daily behaviour.'
>
> (Dennis et al. 1965, 174)

The North Yorkshire mining community which Dennis et al. describe is very similar to that of Neath Port Talbot, where the ideology of the male-dominated culture is still very evident, even amongst the younger generation that have not been a part of the mining way of life. Many other authors (see Massey 1984, Massey and Wainwright 1985, Jackson 1991, Keating 1991) refer to the example of ex-mining communities where daily life for everyone evolved around the shifts in the local mines, not just for the men, but also for the women who were at home washing and cooking around their husbands' and sons' shift patterns. The important acknowledgement of social expectations within economic change was previously noted by Cooke (1983b) and discussed in chapter four.

When considering contemporary society, these stereotypical societal expectations may have been generated through past economic histories, the majority of mines have now closed and economic restructuring has occurred. However, opinions, expectations and behaviours remain in existence as the male-dominated culture has been produced and reproduced through patriarchal conduct in daily routines, to the younger generation. Evidence to support this comes from the research in Neath Port Talbot where the last mine in the area closed in 1988, yet the existence of a male breadwinner society is still evident even amongst the younger generation who have never known mining life. For example the majority of pregnant women in Neath Port Talbot indicated that they had no intention of returning to work after the birth of their child because they felt it was their responsibility to look after their child(ren). However, it is not just in mining communities where a male breadwinner, female homemaker attitude remains. In both West Dorset and Neath Port Talbot the women described how when their partners did fulfil any of the domestic duties, the men referred to the accomplishment as 'help', rather than performing his equal responsibility, inferring that in fact it is the woman's job to carry out all the domestic tasks!

> 'I think it is very hard for men to see you in a different role, other than the one at home. I think my husband tried very hard to be a man of the 90s, but he still says things like "I've washed the dishes". He thinks he has done it for me, and I am constantly battling to say "it isn't my role anymore". He did something for me yesterday; he went to my sister's to borrow the sewing machine. He said to me "you didn't even say thank you", and I felt like saying have you said thank you for all the washing and ironing I have done all weekend. I still do that, I still take responsibility for those things. Sometimes he is home before me and does tea, but he thinks then he has done you a favour, rather than taking responsibility. I find that quite hard. I have spoken to quite a few women who are working and they come up against the same thing and I think it is a hard thing to overcome, they just cannot see it is not your role anymore...The tables have turned and I am the main breadwinner now, and I think I was happy when he was the breadwinner to do all those jobs because I saw it as my role because I was at home. More times than not he is home before me, but he doesn't see it as his role to get tea on. He sometimes does it, but it is a favour.'
>
> (N9, *mother of two, works full-time*)

This perception and reinforcement of roles as male breadwinner, female homemaker, is also backed up by the majority of women in West Dorset who indicated that having a child(ren) had affected the type of job they did. Interestingly, the majority of women in Neath Port Talbot indicated that having a child(ren) had not affected the type of job they do. This is due to the perception of a lack of suitable alternative jobs, in addition to the existence of predetermined societal expectations, which meant that women in Neath Port Talbot have always expected and planned their lives to be dominated by their home lives. Having child(ren) had not affected those plans.

Morris, in her analysis of the locality and the household, discovers what she terms 'local gender ideologies' (1991, 167). This term describes the ideologies that people in the area hold about gender roles and relations, which are embodied in prescriptions about appropriate gender role behaviour. These local gender ideologies were clearly evident within both the case study areas, impacting greatly on women's labour market participation decision making. With reference to Massey's previous work on the South Wales coalfields (1984), Morris makes the claim that although employment for women would clearly have a household impact, the inferior nature of the jobs available would not necessarily be sufficient to make a significant challenge to the gendered perceptions of work roles. This is certainly true of the majority of women in the Neath Port Talbot area who were employed in menial jobs such as shop work, factory work or receptionist in local business, which was chosen because it was employment that they could undertake without it impacting on their home lives.

> 'I wouldn't want to work a full day's work to be honest. This allows me to get on with things like her dance costumes, and get on with things in the house.'
>
> (N2, *dinner lady, mother of one daughter*)

Very few of the women in the Neath Port Talbot area were in career or professional employment. The consequence of this was that the women remained the primary carer for the children, and any employment undertaken must fit around the already established routine with the children's needs, for example being available for the school runs in the morning and afternoon. Evidence of the predominance of past economic histories can also be found from my research in West Dorset. It is an area that was historically dominated by an agricultural farming community. Some of the women I spoke to in the area were either still involved in farming, or talked at length of their parents'/family farms. They described how due to the high demand for manual labour, the woman was almost automatically actively involved in the business, even though sometimes that labour went unnoticed in the home rather than actually providing farm labour. The expectation that woman would contribute to the labour in the household has continued through generations, contributing to the pressure on women in the area to work even though many of them have now moved away from farming lifestyles.

This economic history of the area became clearly evident during my discussions with women regarding the expectations they felt regarding participation in work. For instance, the women who were actively involved in the farming lifestyle (D19 was a farmer, D3 and D7 were farmer's wives) discussed the importance of being able to 'go out' to work. D3 was employed as a nurse tutor and regularly had to leave the family home for overnight stays away from home. Although the period prior to her departure from the home was frantic with advance preparations of school clothes and meals, D3 revealed how she enjoyed her time away as it was the only time she had time to herself.

> 'Sometimes you really cry out for time on your own, and that is why I like to go away because I get a bit of time with a friend and we have a good chinwag and go out for a meal.'
>
> (D3)

D19 also described how, because she didn't 'go out' to work, her family and friends often called her up asking her favours as they did not have time as they had to go to work, even though D19 was a mother of two, a full-time farmer and also working voluntarily for several community groups.

Paasi (1991) discusses how culture dominates the way in which the social relations of a group are structured and shaped. He believes that belonging to a locality or community is mediated through affiliations with the neighbourhood, and are demonstrated in individual daily routines, which become the social practices through which generational ideology becomes produced and reproduced. Paasi (1991) also supports Morris' statement that where a traditional attitude to gender roles has already been established, peer group pressure to maintain those roles will be greatest in highly developed networks, and more especially so where these networks are single sex. Morris believes that this set of conditions describes a familiar pattern of working class male sociability certainly detectable in South Wales. This was certainly supported by the actions of some of the partners of the women I spoke to, who indicated that they would feel as if they had failed if the woman had to go out to work. They felt that it was their role and responsibility to support the family, allowing the woman to be a full-time mother.

> 'He has always supported me…I didn't have to work. As far as he is concerned, it was his place to support me. He would never say, "well you chose to work so you have to get on with it", but I suppose I always knew really that if I had taken on those hours then it was up to me to work around the children.'
>
> (D6, *working mother of two*)

Cox and Mair (1991) also describe how a locality becomes an agent as 'local' identities and interests can breed within a locality, and then individuals act up to that identity by mobilising locally defined organisations to further their interests in a manner that would not be possible were they to act separately. This acknowledges that local cultures can impact on the way in which individuals live their lives, and the ways in which that pressure is reproduced through generations by societal and cultural processes. Evidence of this is provided when comparing the age of the child when the mother returned to work in the two case study areas. Women in West Dorset returned to work much sooner than women in Neath Port Talbot due to the lifestyle differences in societal expectations in the two areas.

> 'I felt pressure from society, not my husband, but my kids at school and I felt I ought to work. I know that my neighbours have tittle-tattled about my neighbour who doesn't work and she has only one kid, so I thought I don't want people thinking that and seeing my husband working all the hours and getting really tired. I need to pay my way and contribute.'
>
> (D7, *mother of four, classroom assistant*)

This work supports Peck's argument that local labour markets develop distinctive characteristics which are reproduced in the local education systems, family support networks and social service regimes.

> 'The capacity to work…is socially produced and reproduced; processes which tend themselves to be culturally embedded, institutionalised, and locally specific.'
>
> (2000, 142)

It must, however, be remembered that although predetermined societal expectations may exist, 'places, certainly when conceptualised as localities, are of course not internally uncontradictory' (Massey 1991, 276). Morris also believes that 'such phenomena are not fully determining, in that some women will take on paid employment even when confronted with such sentiments. Others may be deterred' (1991, 169). This is certainly supported by the findings of my empirical research, where the presence and impact of the predetermined expectations is made evident through the need for each of these women to justify why they were not adhering to 'the norm'. Women in both Neath Port Talbot and West Dorset repeatedly made reference to what their neighbours and their own mother did.

> 'Down at the school I am looked down upon because I work. It is *that* lot from the estate down there, I am working to pay their benefits, so I don't know why they look down their noses.'
>
> (N8, *working mother of three*)

Having acknowledged that uneven development does in fact make a difference to the operation of social, economic and cultural processes at the local level, the locality debate also focused on the impact of cultural issues. In an echo of Paasi's comments on the impact of generational change, Massey stresses that it is people, not places themselves, which are reactionary or progressive (1991, 278). This too is supported by my findings, as women of the older generation (perhaps with children older than ten years) expressed a stronger pressure to conform to much more traditional stereotypes of male breadwinner, female homemaker, to the extent that they believed there was no other option for them. It is the younger generation of mothers who are aware of the alternatives to these stereotypes. However, being aware sometimes makes the decision harder as women agonised about whether they could or should be doing the alternative. If they return to work they feel as if they are 'breaking the rules' and feelings of guilt make the woman feel the need to justify the decision they have made. I shall now explore how the locality debate dealt with the impact of culture on women's decision-making process.

The Impact of Culture

As previously mentioned (see chapter 4), although the focus of the ESRC funded research on 'locality' was the local economy, its argument was that culture is not purely the product of an economic base or a social structure, but is the means by which people gain their identities and create their social world (Duncan and Savage 1991, 159). Contributors to the locality debate, therefore, attempted to address cultural issues directly and consider the impact they had on the local implementation of economic and social processes. Jackson directly addresses his paper to 'some of the neglected cultural dimensions of the "localities studies" debate' (1991, 216), arguing that cultural studies cannot be a separate issue from economic and social enquiry, but rather should be integrated, allowing for a fuller understanding of the processes operating within the locality (also see Paasi 1991). It is the acceptance of the impact of local cultures on everyday lives which is key to the development of the locality debate as a way of understanding the spatial variation of labour market processes at the local level. Additionally it is also the acceptance of the impact of local cultures on women's everyday lives which is key to gaining a full understanding of their decision-making process regarding labour market participation. In an attempt to define the impact of cultural processes further, Paasi argues that culture is the way the social relations of a group are structured and shaped (1991, 241). In Neath Port Talbot, for instance, the mining culture continues to dominate the local societal relations as the expectation amongst men and women is for the man to be the breadwinner and woman to be the homemaker. Paasi attempts to separate two distinct senses of culture which he argues are often conflated: culture as the meaning or values of groups of people; and culture as the institutionalised processes which occur outside of one person's frame of reference. Duncan and Savage (1991) describe how in the first instance, culture is linked to 'experience'; in the second to 'representation'. However, I would argue that our social expectation is formed through our experiences and, therefore, the two are ultimately tied up with each other and it would therefore be impossible to separate them in a distinct manner. Evidence to support this comes from the fact that unconsciously women's economic activity in both Neath Port Talbot and West Dorset had been impacted by the decision their mother had made regarding labour market participation and the experience they had shared as a child. The situation that these women had experienced, therefore, impacted on the expectations they had of themselves and others, when they became mothers.

Jackson (1991) and Paasi (1991) both believe that culture enriches the locality debate by providing an appreciation of the development of the locality, and consequently a further understanding of the daily routines of individuals and the attitudes and actions of institutions. They believe that culture enriches, rather than invalidates, existing approaches in urban and regional studies (Duncan and Savage 1991, 161). It is the recognition of the importance of history to an area's identity and an understanding of the development of its culture, which helps to provide for a fuller understanding of the social processes and structures in the contemporary locality.

At the time of the ESRC's commissioned projects in the mid 1980s (see chapter 4), Britain had experienced a major economic recession as a result of economic restructuring. It was realised that previous labour market analysis was, therefore, difficult to apply. Additionally Massey points out that some of the political debates were being conducted solely at the national level, and, therefore, some of the conclusions drawn from them were unsubstantiated in any rigorous sense by issues of spatial variation. For Massey, then,

> '...the very fact that the national structural changes themselves involved a geographical restructuring meant that people in different parts of the country were experiencing highly contrasting shifts, and that even the trajectories of change could be quite different in one place from in another...it is not simply final outcomes but processes

of change which are significant to people's experience of their
world, this meant that the political implications of these 'structural
changes' were likely to be highly contrasting between one place and
another. Moreover, the spatial variation was reinforced by the fact
that people in different parts of the country had distinct traditions
and resources to draw on in their interpretation of, and their
response to, these changes.'

(Massey 1991, 268)

Massey also points out that conclusions drawn at the national level about policy
implications and changes in political strategy could not be assumed to be
universally applicable, due to the existence of particular traditions and various
social processes in operation in different parts of the country (see chapter 5 for a
more detailed discussion of this). The study of localities provided a recognition
and understanding of the reality and conditions of diversity, and of the actual
interdependent processes which link the local particularities (Massey 1983). The
locality debate came to realise that localities are not simply spatial areas you can
easily draw a line around; they must be defined in terms of the interaction of social
relations or processes. Localities are, therefore, sets of relations which interact in
different ways in different places, and it is therefore necessary to see locality as an
interconnection of processes. This has implications for local labour market theory.

Labour and Locality as Interconnecting Processes

I have argued that the social and cultural processes in operation within each
locality have played a major role on the way in which each woman makes her
decision regarding her participation in the formal labour market. Many authors
within the locality debate argue that a locality cannot be analysed in isolation as a
bounded place. As Cox and Mair argue,

'The locality is embedded in scale and spatial divisions of labour, which
means that each local actor is also linked, whether directly or only
indirectly, to actors outside the locality.'

(Cox and Mair 1991, 201)

Bauder echoes Cox and Mair by acknowledging the influence of external social
factors on shaping employment relationships, although he is keen to point out that
cultural experiences of place and 'labour market segmentation should not be
expressed as elements of an asymmetrical cause-effect model but as mutually
constitutive processes' (2001, 44) which interconnect. According to Bauder, the
crux of spatial segmentation theory is that local labour markets operate 'in different
ways at different times and in different places' (Peck 1996a, 94), and it is the
understanding of locally sensitive interconnecting interdependencies which must
be accounted for, to bring segmentation theory alongside the locality debate in
order to provide a fuller understanding of women's labour market participation.
As Savage et al. point out, it is only through gaining an understanding of the

interlocking interdependencies within each locality that we can truly appreciate the reasons individual actors make the decisions they do. They argue that,

> '...there is very much a greater specificity to individual places, as the connections of different locally derived social entities are likely to be highly specific to each place. This is because the combination and overlapping of local processes produces a locally distinct context for action; this may alter how people think about their world and what is possible in it.'
>
> (Savage et al. 1987, 32)

Savage et al. (1987) acknowledge that spatial variation occurs at the local level, which impacts directly on the lives of the inhabitants in that locality. The combination of locally determined interlocking interdependencies and locally variant expectations within each locality, affects the ways in which people think about their individual lives and what ambitions they believe to be realistic and achievable within it. Certainly the descriptions of their upbringing and the daily lives of the women I spoke to in Neath Port Talbot and West Dorset demonstrate support for the argument that it is only through gaining an understanding of each of the economic, cultural and social interlocking interdependencies within a locality that we can begin to fully appreciate women's decision-making process with regard to entering, or not entering, the formal labour market.

When considering exactly which interdependencies play an active part in women's decision-making process regarding their labour market participation, figure 8.1 provides a framework of the locally dependent processes. Figure 8.1 is compiled from factors identified as significant through a broad reading of locally sensitive labour market theory, the findings from an extensive survey of women's labour market activity in two local authority areas, followed up by intensive in-depth interviews with women about their decision-making process. It can be argued that the proportion of women who are economically active within each local labour market will vary through time and across space according to four factors, which encompass a combination of locally contingent social, cultural and economic influences: employer practices, labour market structures, family structures and societal pressure.

Figure 8.1 Processes shaping women's participation in the local labour market

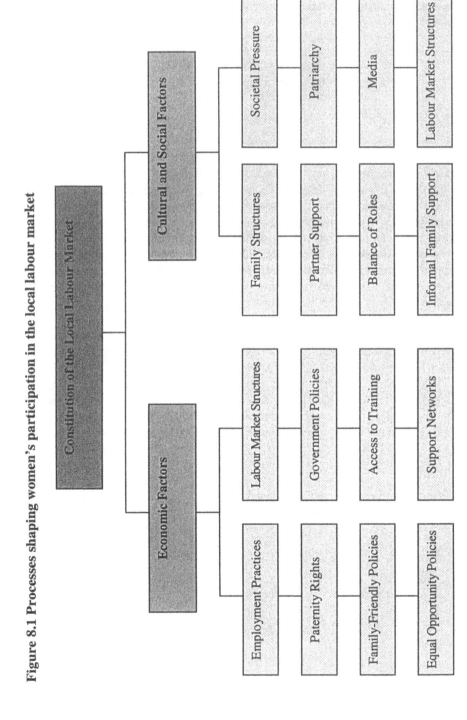

Figure 8.1 outlines the many factors that can actively influence the uneven development of women's participation and progress within the labour market. Each of the factors will have a different role and level of importance for individual women which will vary from place to place, and the combination of which constitute the creation of a local labour market. Firstly, with regard to economic factors, employer practices in the area, including the implementation of family-friendly policies can have an impact on whether or not women enter the labour market. As described in chapter 6, the attitude which an employer adopts towards parental needs, and the implementation of flexible working arrangements (such as those described in figure 3.4) can make it easier for parents, both mother and father, to balance paid work with looking after children or caring for other members of the family, and will, therefore, influence whether or not a woman is able to participate in paid employment (see Craig et al. 1985, Peck 1989b, Mattingly, Hanson and Pratt 1998). The correct implementation of equal opportunity policies is also vital, as although often policies are in place to ensure opportunities are made available to all, in some cases women do not receive equal treatment due to the expectation that they will eventually have children and therefore be less dedicated to their career and subsequently their employer. All too often the responsibility is left with the direct line manager. This can be particularly true for women who try to enter into senior positions, who may come across a 'glass ceiling' (see Davidson and Cooper 1992).

Secondly, labour market structures relate to institutional behaviours within the labour market (see McDowell 2001b, Wilson 1999), which includes the local provision of access to training, job opportunities and support networks such as childcare facilities. Although national governmental policies are vital in encouraging local level institutions to think about issues, it is the provision and support of local initiatives which are the power behind the catalyst to change (see Odland and Ellis 1998). Such local initiatives may provide grants and provision for childcare, or specifically encourage women returners back into the labour market by providing the training necessary for women to update their skills, such as Essex Women Returners' Network which negotiates between individual women and colleges to ensure that all the barriers are taken care of, in order to allow the woman to attend a course. I also support Bennett and McCoshan (1993) who claim that institutional development at present requires many gaps to be filled between national policy and local and individual needs. They argue that gaps lead to system failures because the needs of one sector of activity are not fully registered and responded to in other sectors. This is particularly true of education and training as national targets are set irrespective of the current local needs and demands. Bennett and McCoshan believe that many of these gaps can be overcome only by drawing upon local institutions: they cannot rely solely on national programmes for the specific needs of individuals vary greatly and will not be met under national targets.

'Institutional development is simultaneously a national and a local problem, for which local leadership and local experiences offer national lessons and solutions.'

(Bennett and McCoshan 1993, 289-290)

According to Duncan and Goodwin there now exists a huge volume of literature concerned with variations in local authority service provision, which provides enough documentation for us to accept that these differences are widespread, significant and important socially and politically (1988, 10). Although this is obviously a vital part of the jigsaw in examining barriers preventing women being able to make a genuine choice regarding their labour market participation, the provision of services is not the sole consideration. As described in chapter 6, many women are actually aware of childcare facilities or childminders nearby that are available, but firstly they do not want to use them, and secondly, if they are considering entering into formal employment, more often than not, the demand is for childcare provision to be available outside of the normal 9-5 regime, as described by N18 and D16,

'The hospital had a crèche, but they started at 9am, so that was no good. Also finishing at 9pm, no one is prepared to look after them till that time at night, so that wouldn't work.'

(N18, *mother of two, nurse*)

'It started at 7:30am, but our shifts started at 7:30am, so there was no way you could drop him off and get into the ward.'

(D16, *talking of a crèche at work*)

The problem, therefore, is not so much the availability of childcare facilities, rather the flexibility in hours of provision. It is, therefore, important to acknowledge this when trying to find an explanation for the geographies of women's participation in the labour market.

Economically the two case study areas are very different. Neath Port Talbot is an ex-mining and industrial area with 32 per cent of jobs still in manufacturing. This means that many of jobs available involve shift working in the factories, which often cover unsociable hours. In contrast, West Dorset is a tourist area with an affluent economy, with the service sector as the main employer, which provides opportunities for part-time and/or seasonal employment. Although at the outset, this may appear to provide an explanation for women's economic activity in these areas, however, I argue that, exploring women's decision-making processes and practices raises other important issues. Women in Neath Port Talbot and West Dorset have similar access to (re)training with college courses often provided free during school hours, so that women may attend without an impact being made upon their daily routine. Similarly, as previously mentioned, access to formal childcare facilities raised the same issues, although a few women felt that the availability and cost was a problem, the majority of women argued for the need for more flexibility outside of the 9 to 5 regime. Women in both areas feel that Government policy is set out to encourage women into the workforce, rather than providing 'choices'. This demonstrates that, although economic factors are important as a basis for women's decision making, the local economy itself does not provide the sole explanation for women's labour market participation.

With reference to social and cultural influences, the third main factor which influences the construction of the local labour markets is the presence of household interdependencies (see Granovetter and Tilly 1988, Hanson and Pratt 1988a, 1995, Pratt and Hanson 1991), personal networks (see Tilly 1988) and the presence of family structures (see Garnsey et al. 1985, Peck 1996). By this, I am referring to the level of support offered to a woman by her partner, and other surrounding family members, which can influence whether or not she is able to enter the labour market. Although I have called this category 'family structures', it also includes any close friends and neighbours who help to share the responsibility of parenthood.   As described in chapter 6, household interdependencies and the extent to which domestic responsibilities, household chores and parental care are shared amongst partners will influence the amount of time a mother is able to dedicate to paid employment and can, therefore, as a knock-on effect, influence the type of work which she is able to undertake (see Freeman 1982, Mackenzie 1989a, Zelinsky et al. 1982).   For example, if the mother is, and remains, primarily responsible for all domestic chores and the primary carer for the child(ren), then she may only be able to take on part-time employment, which can be fitted around school times. However, if all the domestic and parental responsibilities are shared equally, the mother may be able to balance a full-time career. Also under the category of 'family structures' we need to examine the influence of generational expectation, by which I am referring to the impact which the action and expectations of previous generations have on the decision-making process of women today (see Chodorow 1992). Is there a connection between whether the woman's mother worked whist raising her children, or did she have a more 'traditional' view and give up work to be at home to raise her children? To what extent are women influenced by the action and expectations of their mothers? In both Neath Port Talbot and West Dorset, women constantly referred to what decisions they had made in relation to what their mother had done when they were a child. Women in both areas consistently referred to the economic activity or inactivity of their friends and neighbours as a means of justification for their decisions.   Often the connection between experience and expectation was an unconscious one, but the connection was definitely evident.

Finally, the existence of predetermined societal roles, that of male breadwinner and female homemaker, will be important (see Rubery 1994), either through generational expectation, the media or patriarchal labour market structures (for 'gender role constraint' see Cockburn 1988, Tivers 1977; for patriarchy see Benn 1999, Hartmann 1979, Holloway 1999, Walby 1986, 1997; for the use of the media see Benn 1999). The regional 'gender contract' (Fosberg 1998) for women can vary from place to place, with the focus either primarily as a worker or as a homemaker. I have argued that it is important therefore to explore how the local labour market structures adapt to the implementation of national strategies (see

Peck 1994). How do individual women manoeuvre between the two roles demanded of them in our new knowledge-based economy; that of a valued employee within a competitive economy and of full-time carer and mother? Are there local initiatives to encourage women into the workforce in some localities, and initiatives to encourage mothers to be at home with their children in others? What are women's perceptions of the opportunities open to them, and the barriers and constraints that might stop them? I argue that the answer to all of these questions, and many more, will only be provided by understanding the interlocking systems of difference, many of which are locally contingent, therefore, as argued in chapter 4, a simple acceptance of nationally aggregated data leads to an ignorance of the 'real' picture.

The social make up of the two case study areas are very different. In Neath Port Talbot the women feel that societal expectation is for them to be at home with the children, and any employment they do undertake must fit in with and not impact upon the domestic responsibilities. Patriarchal structures exist within daily life (as described in chapters 6 and 7), which reinforces the male breadwinner, female homemaker perceptions. In contrast women in West Dorset feel that they are expected to be at work at the same time as looking after the family and so in the majority of cases, the partner has had to take on more of the domestic responsibilities: this may have evened out the gender roles, but these are still far from equal! Women in West Dorset were more aware of the media portrayal for women to 'have it all' by maintaining a career at the same time as experiencing motherhood, and felt that the Government was to blame for the creation of that pressure. Social networks of neighbours and family provide important support structures for women in both Neath Port Talbot and West Dorset. Many women felt that they would have found it impossible to cope with all the demands placed upon them if they had not had informal networks to fall back on for help and support.

In order to gain a full explanation for women's (non)participation in the labour market, I have argued that it is the combination of both economic and social factors, which need to be explored at the local level in order to gain an understanding of women's decision-making process on entering, or not entering, the formal labour market. My research supports Forsberg et al. who believe that 'the importance of historical factors, culture, tradition and specificity of the economy in the place where the individual lives and works, is highlighted in a regional analysis and adds further to the explanation of gender differences' (2000, 28). Regional and local variation cannot, therefore, be ignored, and must be investigated fully if national policy is to address the needs of all women and achieve the desired result of providing 'choice' for individuals. In order to achieve this, I argue that it is necessary to investigate the household interdependencies of women at the individual level, rather than accepting regionally aggregated data as representative. It is only by understanding the complex ways in which individual women negotiate their local networks, that we can begin to find an explanation for the variation of women's participation in the labour market.

I have argued that it is the acknowledgment of interlocking processes both within the individual locality and the connections with wider national processes, and the acceptance of the impact of local cultures on an individual's decision making, that advances the locality debate to help us gain a fuller understanding of women's decision-making process regarding labour market participation. Cox and Mair discuss how the insertion of localised social structures in wider spatial divisions of labour means that place-bound localities can be used to study the connections and reactions to wider regional, national and global processes, for few issues can be exclusively local. Although I agree with Cox and Mair, I would like to take that argument one step further and argue that even within a single locality, interlocking interdependencies combine in different ways for different women. Many of the discussions of culture and locality discuss the idea of belonging to a locality or community albeit at various scales, for example a friendship or neighbourhood (see Cohen 1982) the workplace culture (see McDowell 1997, Deal 1988) or the local labour market area with distinctive daily routines (see Morris 1991). As discussed earlier, when belonging to a community, an individual is often influenced by the activity and opinions of other members. This predetermined societal expectation creates pressure and impacts on the ways in which you negotiate your daily life. Women in West Dorset feel that it is expected of them to be in work, and many of them have been career orientated, received higher education and entered into professional employment. For this reason, women in West Dorset are significantly older than women in Neath Port Talbot when they have had their child(ren). In contrast women in Neath Port Talbot felt that getting married and having children is their primary role, and so settle into this lifestyle younger.

The Public Policy Implications of the Research

Having identified that social and cultural processes have just as much of an impact on labour market participation as economic processes, and explored through an understanding of locality, how 'space makes a difference', I would like to consider the relevance of my findings for labour market policy debates.

There has recently been a re-engagement of academic literature with the policy relevance debate initially raised by Harvey and others during the 1970s (see Banks and MacKian 2000, Massey 2000, Peck 1999, Pollard et al. 2000). Peck (1999), Massey (2001) and Martin (2001) are concerned that despite the vast amount of research that is exploring the geographies of everyday lives, gaining a unique insight into 'real world practices' (Peck 1999, 131), most of this remains disconnected from an examination of the public policy implications. After exploring both national government policy and economic and social processes at the local level, I have to agree with Martin (2001) who questions the forever expanding body of work which seeks to expand and deepen our understanding of society, the economy and the environment, if this knowledge and understanding is not subsequently used to produce a better society. McKie et al. pointed out that the key outcomes of government's recent consultation exercise was that policies and mechanisms were required to achieve a better balance between home life and employment (2001, 235). Having previously explored the key national policies which have been implemented to address the needs of women (see chapter 3), there are three main policy relevant issues that have emerged from my research that I would like to highlight. Firstly, the understanding that social and cultural factors such as predetermined societal expectation and the partner's opinion of gender roles has as much influence on women's decision to work or not work as practical barriers such as childcare, or economic factors such as wanting more money. It should be recognised that the complex interlocking of social interdependencies are influential on women's labour market participation, more so than currently allowed for by academic debate or government policy. Secondly, it is important to understand that even if not initially recognised by the women themselves, generational influences were also important in determining labour market participation. This was made evident by the fact that many women used the example of what their mother did as a means of justifying their own labour market position. Thirdly, in support of the development of fourth generation segmentation theory and the locality debate, my research has demonstrated that some of the factors which influence women (re)entering the formal employment are place-specific and are influenced by the local labour market. Does government policy recognise these factors?

During the process of undertaking this research there have been some important changes in national Government policy that is designed to tackle some of the issues which I have highlighted for working parents. The Government passed a new Employment Act (HM Government 2001) which provides reforms on maternity entitlements, paternity entitlements and flexible working rights. The Government wants to help working mothers to build up a relationship with their child before having to return to work, and so as of 6th April 2003, a mother's maternity leave was increased from 18 weeks to 26 weeks with an increase in the statutory maternity pay from £75 to £100. Additionally there is a right for mothers to take a further 26 weeks unpaid maternity leave. Although in theory this new legislation allows women the right to be at home with their child(ren) for up to a year after the birth, many women may want to go back to work after the initial 26 weeks due to the loss of money if they stay at home. The Government also wants to recognise the role of fathers in caring for their new child(ren) and the need to allow them to be supportive to their partners, and so the new Employment Act will allow fathers the right to two weeks paternity leave to be taken up to 56 days after the birth of the child. Additionally, in order to provide a continuing amount of support for working parents, the Employment Act allows parents (both mother and father) the right to ask for flexible working arrangements. Although this also allows an employer to refuse the request for flexible working, the employer must provide a clear business case for their refusal. Although these developments in maternity and paternity rights are a big step forward toward giving parents more *choices* when they have children, they will only help those women who have chosen to combine motherhood at the same time as maintaining their career. These changes will not affect the cultural predetermined and embedded societal expectations that exist within Britain's localities. Due to the way in which the current central government operates, new policies and Government advice tends to come as a result of work carried out by one government department, rather than joined-up Government initiatives that address the needs of a larger proportion of the population. This book has pointed out that dealing with one policy issue at a time will not suddenly enable lots of women to flood back into the labour market. Evidence of this is provided by the partnerships formed to provide more accessible and more affordable childcare. Although access to childcare facilities is improving, women are not flooding back into the labour market. Joined-up labour market governance is needed between and across different policy fields and spatial scales to increase the participation of individuals in the new economy. In order to provide all women with genuine choices regarding their labour market activity, the government departments need to work together to collaborate research and ideas. For example, Work: Life Balance in DfES and the Women and Equality Unit in DTI could join up to address issues simultaneously in order to provide realistic options and genuine choices for women.

In light of these policy reforms and the findings from my research, I feel there are two main issues which have arisen that need to be acknowledged and considered more carefully by policy makers in Britain. Firstly, the impact of social and cultural processes on women's decision-making process in addition to economic foundations and secondly, the need for national policy to accommodate local conditions. New initiatives such as the longer maternity and paternity allowances and tax credit proposals are still addressing national level economic issues. It may be argued that allowing parents longer amounts of time at home after the birth of a child helps to address some of the concerning issues raised by parents such as the father not being given the same rights as the mother to get involved with the child or the father being allowed the time off to support the mother as she cares for the baby in the first few months. However, the new allowances are dependent on the economic factor of being able to accept the reduced income during a time when you most need the money. Local social and economic contexts are therefore key factors which need to be considered in the formation of national policy as Martin argues, 'local context matters in the formation and practice of policy' (2001, 204), and so locally sensitive and locally derived initiatives, where proposals are worked at the local level and then supported from national funds need to be encouraged.

Although it is recognised that a national regulatory framework is essential to ensure a minimum standard, a partnership between the local council and local businesses could propose an initiative within the area, which if it helps to provide mothers with more *choices* can then be promoted as a 'best practice' example to other local authority areas. Current cross-departmental initiatives, such as Neighbourhood Renewal in England and Communities First in Wales, are demonstrating encouraging signs of effective partnerships between national and local governments who are working together to improve local conditions and meet the demands of local communities. Although this is possibly the only way to allow for a bottom-up approach, we need to be cautious to avoid a situation, as with City Challenge for instance, where areas were competing against each other for scarce government funds (see Byrne 1997). However, it is the spatial targeting of policy and initiatives that is required if we are to move forward to enable mothers to chose the best decision for them and their children.

Conclusion

> 'General processes will always be constituted in particular places and at particular times where they will inevitably be affected by this contingency effect. In this way nearly all processes and forms will be spatially differentiated. This also means that most processes and forms will have some sort of local expression, and very often this local expression will be in a subnational scale.'
>
> (Duncan 1989, 233)

I have demonstrated that when exploring the issues that impact on a woman's decision-making process regarding whether to (re)enter or not (re)enter the labour market after having child(ren), the barriers which prevent a woman from having a genuine choice stem from *interlocking* interdependencies of economic, social and cultural issues.   The aim of my research was to explore the geographies of women's participation in the labour market.   In order to do this I have engaged with contemporary academic debates on labour market theory, feminist theory and locality. Having provided evidence that nationally aggregated data conceals what is actually occurring at smaller scales of analysis. I have explored economic, social and cultural processes at the local level and concluded that spatial variation does make a difference to the ways in which national policies and processes are implemented at the local level. A government is, therefore, not going to be able to help women (re)enter the formal labour market by tackling single issues. In order to gain a fuller understanding of interdependencies to be able to offer genuine choices, different government departments need to collaborate to offer and promote initiatives. Secondly, policy also needs to be aware of and allow for local spatial variation in economic, social and cultural processes, which will affect the way in which a national policy is implemented and received at the local level.   It is, therefore, pertinent that both academics and policy makers acknowledge the varying ways in which the interlocking of economic, cultural and social interdependencies, combine to influence a mother's decision-making process about whether to work or not work once she has had children.

Bibliography

Arber, S. and Gilbert, N. (1982) *Women and Working Lives* London, Macmillan.

Ashton, D. and Maguire, M. (1984) 'Dual labour market theory and the organisation of labour markets' *International Journal of Social Economics* 11(7) 106-120.

Banks, M. and MacKian, S. (2000) 'Jump in! The water's warm: a comment on Peck's "grey geography"' *Transactions of the Institute of British Geographers* 25(2) 249-254.

Barron, R. and Norris, G. (1976) 'Sexual divisions and the dual labour market' in Leonard Barker, D. and Allen, S. (eds) *Dependence and Exploitation in Work and Marriage* London, Longman 47-69.

Bauder, H. (2001) 'Culture in the labour market: segmentation theory and perspectives of place' *Progress in Human Geography* 25(1) 37-52.

Baxter, J. and Eyles, J. (1997) 'Evaluating qualitative research in social geography: establishing "rigour" in interview analysis' *Transactions of the Institute of British Geographers* 22(4) 505-525.

Beechey, V. (1986) 'Women and employment in contemporary Britain' in Beechey, V. (ed) *Women in Britain Today* Milton Keynes, Open University Press, 77-131.

Beechey, V. and Perkins, T. (1985) 'Conceptualising part-time work' in Roberts, B., Finnegan, R. and Gallie, D. (eds) *New Approaches to Economic Life: Economic restructuring: unemployment and the social division of labour* Manchester, Manchester University Press, 246-263.

Beere, C. (1979) *Women and Women's Issues: A handbook of tests and measures* San Francisco, Jossey-Bass.

Benn, M. (1999) *Madonna and Child: Towards a new politics of motherhood* London, Vintage.

Bennett, R. and McCoshan, A. (1993) *Enterprise and Human Resource Development: Local capacity building* London, Paul Chapman.

Bower, C. (2001) 'Trends in female employment' in *Labour Market Trends* February.

Bowlby, S., Lewis, J., McDowell, L. and Foord, J. (1989) 'The geography of gender' in Peet, R. and Thrift, N. (eds) *New Models in Geography Vol. 2* London, Unwin Hyman, 157-175.

Bradley, H. (1994) 'Gendered jobs and social inequality' in *The Polity Reader in Gender Studies* Cambridge, Polity Press 150-158.

Brannen, J. and Moss, P. (1991) *Managing Mothers: Dual career households after maternity leave* London, Unwin Hyman.

Braverman, H. (1974) *Labour and Monopoly Capital: The degradation of work in the twentieth century* London, Monthly Review Press.

Britton, L. (2001) *Family Friendly Policies Needed* Centre for Economic and Social Inclusion, London.

Bryson, C., Budd, T., Lewis, J., and Elam, G. (1999) *Women's Attitudes to Combining Paid Work and Family Life* The Women's Unit, The Cabinet Office.

Byrne, D. (1997) 'National social policy in the United Kingdom' in Pacione, M., *Britain's Cities: Geographies of division in urban Britain* London, Routledge, pp.108-127.

Callender, C., Millward, N., Lissenburgh, S. and Forth, J. (1996) *Maternity Rights and Benefits in Britain* Department for Social Services, Research report 67.

Campbell Clark, S. (2000) *Work Cultures and Work/Family Balance* Work and Family: Expanding Horizons Conference paper, San Francisco, March.

Castells, M. (1978) *City, Class and Power* London, Macmillan.

Castro, A., Mehaut, P., and Rubery, J. (eds) (1992) *International Integration and Labour Market Organisation* London, Academic Press.

Chodorow, N. (1992) *Knowing Women:Feminism and knowledge* Oxford, Blackwell.

Cockburn, C. (1988) 'The gendering of jobs: workplace relations and the reproduction of sex segregation' in Walby, S. (ed) *Gender Segregation at Work* Milton Keynes, Open University Press, 29-54.

Cohen, A. (ed) (1982) *Belonging* Manchester, Manchester University Press.

Cook, I. and Crang, P. (1995) *Doing Ethnographies* London, Institute of British Geographers.

Cooke, P. (1981) 'Tertiarisation and socio-spatial differentiation in Wales' *Geoforum* 12(4) 319-330.

Cooke, P. (1983a) 'Labour market discontinuity and spatial development' *Progress in Human Geography* 7(4) 543-565.

Cooke, P. (1983b) *Theories of Planning and Spatial Development* London, Hutchinson and Co.

Coote, A. and Campbell, B. (1982) *Sweet Freedom: The struggle for women's liberation* Oxford, Blackwell.

Cox, K. and Mair, A. (1989) 'Levels of abstraction in locality studies' *Antipode* 21(2) 121-132.

Cox, K. and Mair, A. (1991) 'From localised social structures to localities as agents' *Environment and Planning A* 23(2) 197-213.

Craig, C., Garnsey, E. and Rubery, J. (1985) 'Labour market segmentation and women's employment: a case study of the United Kingdom' *International Labour Review* 124(3) 267-280.

Crang, M (2001) 'Field work: making sense of group interviews' in Dwyer, C. and Limb, M. (eds) *Qualitative Methodologies for Geographers* London, Arnold.

Crompton, R. and Harris, F. (1998) 'Explaining women's employment patterns: orientations to work revisited' *British Journal of Sociology* 49(1) 118-136.

Crompton, R., Harris, L. and Walters, P. (1990) 'Gender relations and employment' *British Journal of Sociology* 41(3) 329-350.

Crompton, R. and Mann, M. (eds) (1986) *Gender and Stratification* Cambridge, Polity.

Currell, M. (1974) *Political Women* London, Croom Helm.

Davidson, M. and Cooper, C. (1992) *Shattering the Glass Ceiling: The woman manager* London, Paul Chapman.

Davis, K. (1988) 'Wives and work: a theory of the sex role revolution and its consequences' in Dornbusch, S. and Stoker, M. (eds) *Feminist, Children and New Families* New York, Guilford Press.

Davis, R. (1975) *Women and Work* London, Hutchinson.

Daycare Trust, The (2002) *The Price Parents Pay* http://www.daycaretrust.org.uk accessed March 2002.

Deal, T. (1988) *Corporate Cultures: The rites and rituals of corporate life* London, Penguin.

Dennis, N., Henriques, F. and Slaughter, C. (1956) *Coal is Our Life: An analysis of a Yorkshire mining community* London, Eyre and Spottiswoode.

Deseran, E., Li, J. and Wojtkiewicz, R. (1993) 'Household structure, labour market characteristics and female labour force participation' in Singelmann, J. and Deseran, F. (eds) *Inequalities in Labour Market Areas* Colombia, Westview Press, 165-190.

Dex, S. (1986) *Women's Occupational Mobility: A lifetime perspective* London, Macmillan.

Dex, S., Clark, A. and Taylor, M. (1995) *Household Labour Supply* London, Department of Employment.

DfEE (2000b) *Changing Patterns in a Changing World* Department for Education and Employment.

Doeringer, P. and Piore, M. (1971) *Internal Labour Markets and Manpower Analysis* Lexington, DC Health.

Dorset County Council (2002) *The Dorset Data Book: Data and statistics for the County of Dorset*

Dti (2000) *Work and Parents: Competitiveness and choice: A green paper* Department of Trade and Industry.

Dti (2001) *Government Response to the Recommendations from the Work and Parents Taskforce* Department of Trade and Industry.

Du Bois, B. (1983) 'Passionate scholarship: notes on values knowing and method in feminist research' in Bowles, G. and Duelli Klein, R. (eds) *Theories of Women's Studies* London, Routledge and Kegan Paul.

Duncan, J. and Duncan, N. (2001) 'Theory in the field' *Geographical Review* 91(1/2) 399-406.

Duncan, S. (1989) 'What is locality?' in Peet, R. and Thrift, N. (eds) *New Models in Geography Vol. 2* London, Unwin Hyman, 21-252.

Duncan, S. (2000) 'Introduction: theorising comparative gender inequality' in Duncan, S. and Pfau-Effinger, B. (eds) *Gender, Economy and Culture in the European Union* London, Routledge.

Duncan, S. and Edwards, R. (1999) *Lone Mothers Paid Work and Gendered Moral Rationalities* Basingstoke, Macmillan.

Duncan, S. and Goodwin, M. (1988) *The Local State and Uneven Development* Cambridge, Polity.

Duncan, S. and Savage, M. (1991) 'New perspectives on the locality debate' *Environment and Planning A* 23(2) 156-164.

Duncan, S. and Smith, D. (2002) 'Family Geographies and Gender Cultures' *Social Policy and Society* 1(1) 21-34.

EOC (2000a) *Women and Men in Britain: The work-life balance*, Manchester, Equal Opportunities Commission.

EOC (2000b) *Women and Men in Britain: At the millennium*, Manchester, Equal Opportunities Commission.

Epstein Jayaratne, T. (1997) 'The value of quantitative methodology for feminist research' in Hammersley, M. (ed) *Social Research: Philosophy, Politics and Practice* London, Sage 109-123.

Esping-Andersen, G. (1990) *The Three Worlds of Welfare Capitalism* Cambridge, Polity.

Esping-Andersen, G. (2000) 'Notes and issues: Interview on postindustrialism and the future of the welfare state' *Work, Employment and Society* 14(4) 757-769.

Eyles, J. and Donovan, J. (1990) *The Social Effects of Health Policy* Aldershot, Avebury Press.

Finch, J. (1984) '"It's great to have someone to talk to": the ethics of interviewing women' in Bell, C. and Roberts, H. (eds) *Social Researching: Problems, Politics and Practice* London, Routledge and Keegan Paul, 70-89.

Flametree (2001) *Work:Life Balance* www.flametree.co.uk accessed 5th February 2002.

Forsberg, G. (1998) 'Regional variations in gender contract: gendered relations in labour markets, local politics and everyday life in Swedish regions' *Innovation* 11(2) 191-210.

Forsberg, G., Gonas, L. and Perrons, D. (2000) 'Paid work: participation, inclusion and liberation' in Duncan, S. and Pfau-Effinger, B. (eds) *Gender, Economy and Culture in the European Union* London, Routledge.

Freeman, C. (1982) 'The "Understanding Employer" in West, J. (ed) *Work, women and the Labour Market* London, Routledge and Kegan Paul, 135-153.

Friedman, A. (1977) *Industry and Labour: Class struggle at work and monopoly capitalism* London, Macmillan.

Gerson, K. (1985) *Hard Choices: How women decide about work, career and motherhood* London, University of California Press.

Gerson, K. (2002) 'Moral dilemmas, moral strategies, and the transformation of gender: lessons from two generations of work and family change' *Gender and Society* 16 (1) 8-28.

Giddens, A. (1998) *The Third Way: The renewal of social democracy* Cambridge, Polity.

Gillham, B. (2000) *Case Study Research Methods* London, Continuum.

Ginn, J., Arber, S., Brannen, J., Dale, A., Dex, S., Elias, P., Moss, P., Pahl, J., Roberts, C. and Rubery, J. (1996) 'Feminist fallacies: a reply to Hakim on women's employment' *British Journal of Sociology* 47(1) 167-174.

Glaser, B. (1992) *Emergence vs. Forcing: The basics of grounded theory analysis* Mill Valley, CA: Sociology Press.

Gordon, D. (1972) *Theories of Poverty and Underdevelopment* Massachusetts, D.C.Heath.

Gordon, D., Edwards, R. and Reich, M. (1982) *Segmented Work, Divided Workers: The historical transformation of labour in the US* Cambridge, Cambridge University Press.

Granovetter, M. and Tilly, C. (1988) 'Inequality in the labour process' in Smelser, N. (ed) *Handbook of Sociology* California, Saga, 175-221.

Green, A. and Owen, D. (1990) 'The development of a classification of travel-to-work areas' *Progress in Planning* 34(1) 1-92.

Green, A. and Owen, D. (1998) *Where are the Jobless? Changing unemployment and non-employment in cities and regions* Bristol, Policy Press.

Hakim, C. (1979) *Occupational Segregation: A comparative study of the degree and pattern of the differentiation between men and women's work in Britain, the US and other countries* London, Department of Employment.

Hakim, C. (1987) *Homeworking in Wages Council Industries: A study based on wages inspectorate records of pay and earnings* London, Department of Employment, Research Paper 37.

Hakim, C. (1993) 'The myth of rising female employment' *Work, Employment and Society* 7(1) 97-120.

Hakim, C. (1995) 'Five feminist myths about women's employment' *British Journal of Sociology* 46(3) 429-453.

Hakim, C. (1996a) *Key Issues in Women's Work: Female heterogeneity and the polarisation of women's employment* London, Athlone.

Hakim, C. (1996b) 'The sexual division of labour and women's heterogeneity' *British Journal of Sociology* 47, 78-187.

Hakim, C. (2000) *Work-Lifestyle Choices in the 21st Century: Preference theory* Oxford, Oxford University Press.

Hammersley, M. (1995) *The Politics of Social Research* London, Sage.

Hanson, S. (1988) 'Spatial dimensions of the gender division of labour in a local labour market' *Urban Studies* 9, 180-202.

Hanson, S. (1997) 'As the world turns: new horizons in feminist geographical methodologies' in Jones III, J.P., Nast, H. and Roberts, S. (eds) *Thresholds in Feminist Geography* Oxford, Rowman and Littlefield.

Hanson, S. and Johnson, I. (1985) 'Gender differences in work trip length: explanations and implications' *Urban Geography* 6(3) 193-219.

Hanson, S. and Pratt, G. (1988a) 'Reconceptualising the links between home and work in urban geography' *Economic Geography* 64(4) 299-321.

Hanson, S. and Pratt, G. (1988b). 'Spatial dimensions of the general division of labour in a local labour market' *Urban Geography* 9(2) 180-208.

Hanson, S. and Pratt, G. (1992) 'Dynamic dependencies: a geographic investigation of local labour markets' *Economic Geography* 68, 373-405.

Hanson, S. and Pratt, G. (1995) *Gender, Work and Space* London, Routledge.

Harrison, C. and Burgess, J. (1994) 'Social constructions of nature: a case study of conflicts over the development of Rainham Marshes' *Transactions of the Institute of British Geographers* 19(3) 291-310.

Harrison, P. (1983) *Inside the Inner City* Harmondsworth, Penguin.

Hartmann, H. (1979) 'Capitalism, patriarchy and job segregation by sex' in Eisenstein, Z. (ed) *Capitalist Patriarchy and the Case of Social Feminism* New York, Monthly Review Press.

Herod, A. (1993) 'Gender issues in the use of interviewing as a research method' *Professional Geographer* 45(3) 308-317.

Hewitt, P. (2002) *Address to the Daycare Trust Conference*, London 13th November.

HM Government (1998) *Meeting the Childcare Challenge* www.dfes.gov.uk/childcare accessed March 2002.

HM Government (2001) Employment Bill [HC BILL 2001(44)].

Hogarth, T., Hasluck, C., Pierre, G., Winterbotham, M. and Vivan, D. (2000) *Work-Life Balance 2000: Baseline study of work-life practices in Great Britain* Department of Education and Employment.

Hoggart, K., Lees, L. and Davies, A. (2002) *Researching Human Geography* London, Arnold.

Holloway, S. (1999) 'Mother and worker? The negotiation of motherhood and paid employment in two urban neighbourhoods' *Urban Geography* 20(5) 438-460.

Houston, D., Marks, G., Allcock, L., Lloyd, K. and Waumsley, J. (2000) *The Future of Work: Families and work* University of Kent at Canterbury, Work-Life Research Group.

Humphrys, G. (1972) *South Wales* Newton Abbot, David and Charles.

Hyman, H., Cobb, W., Feldman, C., Hart, C. and Stember, C. (1954) *Interviewing in Social Research* Chicago, University of Chicago Press.

Industrial Society (2001) *The Industrial Society: Consultation Response* http://www.indsoc.co.uk accessed March 2002.

Institute of Directors (2001) *Institute of Directors: Employment Comment* http://www.iod.com accessed March 2002.

Jackson, P. (1991) 'Mapping meanings: a cultural critique of locality studies' *Environment and Planning A* 23(2) 215-228.

Jarvis, H. (1991) 'Housing, labour markets and household structure: questioning the role of secondary data analysis on sustaining the polarizing debate' *Regional Studies* 31(5) 521-532.

Jay, M. (2000a) *Women. Future – It's a whole new conversation transforming the world of work* 5th April.

Jay, M. (2000b) *Address to National Family Planning and Parenting Conference* 13th April.

Jay, M. (2001a) *Address to the Annual Women in Business Forum* INTEGRA 29th March.

Jay, M. (2001b) *Address to the Women in Business Forum* Canada High Commission, 8th February.

Johnson-Welch, C. Bonnard, P., Spring Richie, A., Strickland, R. and Sims, M. (2000) *They Can't Do It At All: A call for expanded investments in childcare* International Centre for Research on Women Research Paper.

Joseph, G. (1983) *Women at Work* Oxford, Philip Allan.

Keating, J. (1991) *Counting the Cost: A family in a miners' strike* Barnsley, Wharncliffe Publishing.

Kirkpatrick, J. (1974) *Political Women* New York, Basic Books.

Kreckel, R. (1980) 'Unequal opportunity structure and labour market segmentation' *Sociology* 14(4) 525-550.

Labour Party (1997) *New Labour: Because Britain deserves better* London, Labour Party

Labour Party (2001) *Ambitions for Britain: Labour's manifesto* Surrey, HH Associates.

La Valle, I., Finch, S., Nove, A. and Lewin, C. (1999) *Parents' Demand for Childcare* Department for Education and Employment.

Leckie, G. (1993) 'Female farmers in Canada and gender relations of a restructuring agricultural system' *The Canadian Geographer* 37(3) 212-230.

LeCompt, M. and Schensul, J. (1999) *Analysing and Interpreting Ethnographic Data* Walnut Creek, CA: AltaMira Press.

Little, J., Peake, L. and Richardson, P. (eds) (1988) *Women in Cities* Hampshire, Macmillan.

Luxton, M. (1990) 'Two hands for the clock: changing patterns in the gendered division of labour in the home' in Arat-Koc, S., Luxton, M. and Rosenberg, H. (eds) *Through the Kitchen Window: The politics of home and family* Toronto, Garamond, 39-55.

Mackenzie, S. (1986) 'Women's responses to economic restructuring: changing gender, changing space' in Barrett, M. and Hamilton, R. (eds) *The Politics of Diversity: Feminism, Marxism and Canadian Society* London, Verso, 81-100.

Mackenzie, S. (1988) 'Balancing our space and time: the impact of women's organisation on the British city, 1920-1980' in Little, J., Peake, L. and Richardson, P. (eds) *Women in Cities* Hampshire, Macmillan, 41-60.

Mackenzie, S. (1989a) 'Restructuring the relations of work and life: women as environmental actors, feminism as geographical analysis' in Kobayashi, A. and Mackenzie, S. (eds) *Remaking Human Geography* London, Unwin Hyman, 40-61.

Mackenzie, S. (1989b) 'Women in the City' in Peet, R. and Thrift, N. (eds) *New Models of the City Vol. 2* London, Unwin Hyman, 109-123.

Marks, S., Huston, T., Johnson, E. and MacDermid, S. (2001) 'Role balance among white married couples' *Journal of Marriage and the Family* 63(4) 1083-1098.

Marshall, B. (1994) *Engendering Modernity: Feminism, social theory and social change* Cambridge, Polity.

Martin, J. (1986) 'Returning to work after childbearing: evidence from the Women and Employment Survey' *Population Trends* 43(2) 23-30.

Martin, R. (2001) 'Geography and public policy: the case of the missing agenda' *Progress in Human Geography* 25(2) 189-210.

Massey, D. (1979) 'In what sense is it a regional problem?' *Regional Studies* 13(2) 233-243.

Massey, D. (1980) *Industrial Restructuring as Class Restructuring: Some examples of the implications of industrial change for class structure* Centre for Environmental Studies Working Note 604, London, CES.

Massey, D. (1983) 'The shape of things to come' *Marxism Today* September, 6-13.

Massey, D. (1984) *Spatial Divisions of Labour: Social structures and the geography of production* Basingstoke, Macmillan.

Massey, D. (1991) 'The political place of locality studies' *Environment and Planning A* 23(2), 267-281.

Massey, D. (1994) *Space, Place and Gender* Cambridge, Polity.

Massey, D. (2000) 'Editorial: practising political relevance' *Transactions of the Institute of British Geographers* 25(2) 131-133.

Massey, D. (2001) 'Geography on the agenda' *Progress in Human Geography* 25, 5-17.

Massey, D. and Meegan, R. (1982) *The Anatomy of Job Loss: The how, why and where of employment decline* London, Macmillan.

Massey, D. and Meegan, R. (1985) *Politics and Method: Contrasting studies in industrial geography* London, Methuen.

Massey, D. and Wainwright, H. (1985) 'Beyond the coal fields: the work of the miners' support groups' in Beynon, H. (ed) *Digging Deeper: Issues in the miners' strike* London, Verso, 149-168.

Mattingly, D., Hanson, S. and Pratt, G. (1998) 'Women's lives, local geographies and the effects of maternity breaks on women's employment' *Michigan Feminist Studies* 12, 1-25.

McDowell, L. (1992) 'Doing gender: feminism, feminists and research methods in human geography' *Transactions of the Institute of British Geographers* 17(4) 399-416.

McDowell, L. (1997) *Capital Culture: Gender at work in the city* Oxford, Blackwell.

McDowell, L. (2000) 'Feminists rethink the economic: the economics of gender/the gender of economics' in Clark, G., Feldman, M. and Gertler, M., *The Oxford Handbook of Economic Geography* Oxford University Press, Oxford, 497-517.

McDowell, L. (2001a) 'Father and Ford revisited: gender, class and employment change in the new millennium' *Transactions of the Institute of British Geographers* 24(4) 448-464.

McDowell, L. (2001b) 'Working with young men' *Geographical Review* 91(1/2) 201-214.

McKie, L., Bowlby, S. and Gregory, S. (2001) 'Gender, caring and employment in Britain', *Journal of Social Policy* 30(2) 233-258.

Merrifield, A. (1995) 'Situated knowledge through exploration: reflections on Bunge's "Geographical expectations"' *Antipode* 27(1) 49-70.

Michon, F. (1987) 'Segmentation, employment structures and productive structures' in Tarling, R. (ed) *Flexibility in the Labour Market* London, Academic Press, 23-55.

Mitchell, J. (1983) 'Case and situation analysis' *Sociological Review* 31(2) 187-211.

Morris, L. (1991) 'Locality studies and the household' *Environment and Planning A* 23(2) 165-177.

National Assembly of Wales (1998) *National Childcare Strategy for Wales.*

National Assembly of Wales (2001) *Digest of Wales Local Area Statistics.*

Neath Port Talbot County Borough Council (1999) *Economic Development Strategy 1999-2002.*

Neath Port Talbot County Borough Council (2002) *Economic Development Strategy 2002-2005.*

Oakley, A. (1981) *Subject Woman* Oxford, Martin Robertson.

Odland, J. and Ellis, M. (1998) 'Variations in the labour force experience of women across large metropolitan areas in the United States' *Regional Studies* 32(4), 333-347.

ONS (1997). *Women's Incomes Over a Lifetime* Office of National Statistics.

ONS (2000) *Social Trends 30* Office of National Statistics.

Orenstein, P. (2000) *Women on Work, Love, Children and Life* London, Piatkus.

Paasi, A. (1991) 'Deconstructing regions: notes on the scales of spatial life' *Environment and Planning A* 23(2) 239-256.

Pahl, R. (1985) 'The restructuring of capital, the local political economy and household work strategies' in Gregory, D. and Urry, J. (eds) *Social Relations and Spatial Structures* Basingstoke, Macmillan 242-264.

Palm, R. and Pred, A. (1974) *A Time-Geographic Perspective on Problems of Inequality for Women* Institute of Urban and Regional Development, University of California, Berkeley Working Paper No. 236.

Parents at Work (2001) www.parentsatwork.org.uk accessed 10th February 2002.

Peck, J. (1989) 'Literature surveys: labour market segmentation theory' *Labour and Industry* 2(1) 119-144.

Peck, J. (1989a) 'Reconceptualising the local labour market: space, segmentation and the state' *Progress in Human Geography* 13(1) 42-61.

Peck, J. (1996a) 'Review: Hanson, S. and Pratt, G. 1995 "Gender, Work and Space" London, Routledge' *Antipode* 28(4) 343-45.

Peck, J. (1996b) *Work-Place: The social regulation of labour markets* London, Guilford.

Peck, J. (1999) 'New Labourers? Making a new deal for the workless class' *Environment and Planning C: Government Policy* 17(3) 345-372.

Peck, J. (2000) 'Places of Work' in Sheppard, E. and Barnes, T. (eds) *A Companion to Economic Geography* Oxford, London 113-148.

Pfau-Effinger, B. (1998). 'Gender, cultures and the gender arrangement – a theoretical framework for cross-national gender research' *Innovation* 11(2) 147-166.

Picchio, A. (1992) *Social Reproduction: The political economy of the labour market* Cambridge, Cambridge University Press.

Pollard, J., Henry, N., Bryson, J. and Daniels, P. (2000) 'Shades of grey? Geographers and policy' *Transactions of the Institute of British Geographers* 25(2) 243-248.

Pratt, A. (1991) 'Discourses of locality' *Environment and Planning A* 23(2) 257-266.

Pratt, G. (1993) 'Reflections on post-structuralism and feminist empirics, theory and practice' *Antipode* 25 (1) 51-63.

Pratt, G. (1994) 'Feminist geographies' in Johnston, R., Gregory, D. and Smith, D. (eds) *The Dictionary of Human Geography* Cambridge, Blackwell, 192-196.

Pratt, G. and Hanson, S. (1988) 'Gender, class and space' *Environment and Planning D: Society and space* 6(1) 15-35.

Pratt, G. and Hanson, S. (1991) 'On the links between home and work: family-household strategies in a buoyant labour market' *International Journal of Urban and Regional Research* 15(1) 55-74.

Rahman, M., Palmer, G., Kenway, P. and Howarth, C. (2000) *Monitoring Poverty and Social Exclusion* Joseph Rowntree Foundation.

Rake, K. (2000) *Women's Incomes Over a Lifetime* The Women's Unit, The Cabinet Office.

Reich, M., Gordon, D. and Edwards, R. (1973) *A Theory of Labour Market Segmentation* A.E.R papers and proceedings, 359-65.

Reinharz, S. (1982) 'Experimental analysis: a contribution to a feminist research' in Bowles, G. and Duelli Klein, R. (eds) *Theories in Women's Studies* London, Routledge and Kegan Paul.

Reskin, B. and Roos, P. (1990) *Job Queues, Gender Queues – Explaining women's inroads into male occupations* Philadelphia, Temple University Press.

Richards, W. (1988) *Women and the Labour Market: Gender and disadvantage* Trinity College Dublin.

Riddell (2001) 'Marriage made in Millbank' *The Guardian* 28.

Rose, G. (1993) *Feminism and Geography: The limits of geographical knowledge* Cambridge, Polity Press.

Rose, G. (1995) 'Distance, surface, elsewhere: a feminist critique of the space of phallocentric self knowledge' *Environment and Planning D: Society and space* 13(6) 716-81.

Rose, G. (1997) 'Situating knowledges: positionality, reflexivities and other tactics' *Progress in Human Geography* 21(3) 305-320.

Rubery, J. (1988) 'Employers and the labour market' in Gallie, D. (ed) *Employment in Britain* Oxford, Basil Blackwell, 251-280

Rubery, J. (1992) 'Productive systems, international integration and the single European market' in Castro, A., Mehaut, P. and Rubery, J. (eds) *International Integration and Labour Market Organisation* London, Academic Press 244-261.

Rubery, J. (1994) 'The British production regime: a societal-specific system?' *Economy and Society* 23(3) 335-354.

Rubery, J. and Fagan, C. (1995) 'Comparative industrial relations research towards reversing the gender bias' *British Journal of Industrial Relations* 33(2) 209-236.

Sainsbury, D. (ed) (1994) *Gendering Welfare States* London, Sage.

Savage, M., Barlow, J., Duncan, S. and Saunders, P. (1987) 'Locality research: the Sussex Programme on economic restructuring social change and the locality' *Quarterly Journal of Social Affairs* 3(1) 27-51.

Sayer, A. (1992) *Method in Social Science: A realistic approach* London, Routledge.

Sayer, A. and Walker, R. (1992) *The New Social Economy: Reworking the division of labour* Oxford, Blackwell.

Schoenberger, E. (1992) 'Self-criticism and self-awareness in research: a reply to Linda McDowell' *Professional Geographer* 44(2) 215-218.

Schramm, W. (1971) *Notes on Case Studies of Instrumental Media Projects* Academy for Educational Development, Washington D.C.

Sichterman, B. (1988) 'The conflict between housework and employment: some notes on women's identity' in Jenson, J., Hagen, E. and Reddy, C. (eds) *Feminisation of the Labour Force: Paradoxes and promises* New York, Oxford University Press, 276-287.

Stacey, M. (1969) 'The myth of community studies' *British Journal of Sociology* 20(2) 34-47.

Staeheli, L. and Lawson, V. (1994) 'A discussion of "Women in the field": the politics of feminist fieldwork' *Professional Geographer* 46, 96-102.

Stake, R. (1995) *The Art of Case Study Research: Perspectives on practice* London, Sage.

Thompson, K. (1985) 'The labour process and deskilling' in Thompson, K. (ed) *Work, Employment and Unemployment: Perspectives on work and society* Milton Keynes, Open University Press, 67-86.

Tivers, J. (1977) *Constraints on Urban Activity Patterns of Women with Young Children* London, University of London.

Tivers, J. (1986) *Women Attached: The daily lives of women with young children* London, Croom Helm.

Turner, R. (2000) *Coal Was Our Life* Sheffield, Sheffield Hallam University Press.

United Nations Development Project (1992) *Human Development Report* Oxford, Oxford University Press.

Urry, J. (1981) 'Localities, regions and social class' *International Journal of Urban and Regional Research* 5(4) 455-474.

Villeneuve, P. and Rose, G. (1988) 'Gender and the separation of employment from home in metropolitan Montreal 1971-1981' *Urban Geography* 9(2) 155-179.

Walby, S. (ed) (1986) *Patriarchy at Work: Patriarchal and capitalist relations in employment* Cambridge, Polity.

Walby, S. (1990) *Theorizing Patriarchy* Oxford, Basil Blackwell.

Walby, S. (1997) *Gender Transformations* London, Routledge.

Wallace, M., Charlton, J. and Denham, C. (1995) 'The new OPCS area classifications' *Population Trends* 79, 15-30.

Watson, G. and Fothergill, B. (1993) 'Part-time employment and attitudes to part-time work' *Employment Gazette* 110, 213-220.

Weedon, C. (1987) *Feminist Practice and Poststructuralist Theory* Oxford, Blackell.

West Dorset Development Council (2000a) *Performance Plan.*

West Dorset Development Council (2000b) *Economic Development Strategy.*

West, J. (ed) (1982) *Work, Women and the Labour Market* London, Routledge and Kegan Paul.

WEU (Women and Equality Unit) (2000) *Top 20 Things the Government is Doing for Women.*

WEU (Women and Equality Unit) (2000a) *Women's Script* unpublished, internal document.

WEU (Women and Equality Unit) (2001) *Women and Work: Challenge and opportunity* Cabinet Office.

Wilkinson, F. (1983) 'Productive systems' *Cambridge Journal of Economics* 7(3/4) 413-429.

Wilkinson, H. (1998) 'The family way: navigating a third way in family policy' in Hargreaves, I. and Christie, I. (eds) *Tomorrow's Politics: The third way and beyond* London, DEMOS 111-125.

Wilson, R. (1999) *UK Labour Market Prospects: Review of the economy and employment 1998/9* Institute for Employment Research: Table 1.3.

Work and Parents Taskforce (2001) *About Time: Flexible working* Department of Trade and Industry.

Women and Geography Study Group of the Institute of British Geographers (1984) *Geography and Gender: An introduction to feminist geography* London, Hutchinson.

WU (Women's Unit) (2000a) *More Choice for Women in the New Economy: The facts* The Cabinet Office.

WU (Women's Unit) (2000b) *Voices: The progress report* The Cabinet Office.

Zelinsky, W., Monk, J. and Hanson, S. (1982) 'Women and geography: a review and prospectus' *Progress in Human Geography* 6, 317-366.

Index

Printed and bound by CPI Group (UK) Ltd, Croydon, CR0 4YY

22/10/2024

01777640-0001